Watching Insects

Dr Abraham Verghese
Ph. D., FRES (London)

Watching Insects
Copyright © 2015 Dr Abraham Verghese
All rights reserved

No part of this publication may be reproduced, stored in a retrieval system, or transmitted, in any form or by means electronic, mechanical, photocopying, or otherwise, without prior written permission of the Author.

Requests for permission should be addressed to
Dr Abraham Verghese (watchinginsects@gmail.com).

Contents

	Contents	3
1	Preface	9
2	Preamble	11
3	Insects and Their Interesting Body Parts	15
4	Insect Antennae	22
5	Insect Legs	27
6	How to Watch Insects	32
7	Crawlers and Earth-Bound Insects	33
8	Insects of Grasses and Herbs	34
9	Watching Nocturnal Insects	35
10	Watching Insects Visiting Flowers	36
11	Watching Flying Insects	39
12	Collection and Preservation	41
13	Making of A Good Insect Box	44

14	Insect Box for Storing Insect Specimens	45
15	Watch Grasshoppers and Crickets	47
16	Watch Crickets	52
17	Watch Stick Insect and Leaf Insects	55
18	Watch Earwigs	56
19	Watch Roaches	58
20	Watch Preying Mantids	60
21	Watch Lice	62
22	Watch Bugs	63
23	Watch Assasin Bugs	64
24	Watch Coreids	65
25	Watch Shield Bugs	66
26	Watch Tingids	67
27	Watch Water Bugs	68
28	Watch Cicadas	70
29	Watch Spit Bugs	72
30	Watch Tree Hoppers	73

31	Watch Hoppers, Jumping Lice and Other Plant Lice	74
32	Watch Whiteflies	77
33	Watch Scales and Mealybugs	79
34	Watch Thrips	82
35	Watch Green Lace-Wings	84
36	Watch Ant-Lions	86
37	Watch Butterflies and Moths	88
38	Watch Blues and Sulphurs	92
39	Watch Browns or Satyrids	94
40	Watch Brush-Footeds or Nymphalids	96
41	Watch Lesser Moths	98
42	Watch Bag Worm Moths	100
43	Watch Limacrodids	101
44	Watch Wax Moths	103
45	Watch Grass Moth	104
46	Watch Flour Moth and Pod Moths	105
47	Watch Pyralids	106

48	Watch Plume Moths	107
49	Watch Silk Moths	108
50	Watch Tussock Moths	109
51	Watch Noctuid Moths	110
52	Watch Hawk Moths	111
53	Watch Flies	113
54	Watch Fleas	116
55	Watch Leaf Cutting Bees	117
56	Watch Wasps	118
57	Watch Beetles	120
58	Watch Ground Beetles	121
59	Watch Water Beetles	122
60	Watch Rove Beetles	123
61	Watch Chafers	124
62	Watch Jewel Beetles	126
63	Watch Fireflies	127
64	Watch Click Beetles	128

Watching Insects

65	Watch Lady Bird Beetles	129
66	Watch Blister Beetles	132
67	Watch Leaf Eating Beetles	133
68	Watch Longhorned or Longicorn Beetles	134
69	Watch Pulse Beetles	136
70	Watch Weevils	137
71	Watch Scotylids	139
72	Watch Social Insects	140
73	Watch Bees	141
74	Watch Ants	145
75	Watch Termites	150

1. Preface

One of the pleasures of watching nature is getting to know the various life forms that exist around us. The best way of appreciating the natural history is to watch the most abundant of creatures in the environment and to this category belongs insects which consist of almost two-thirds of all moving animals. Many of them occur very proximal to us enabling closer observation and recording. Barring the exceptions like mosquitoes, flies, cockroaches, bed bugs, head lice and fleas a vast majority of insects are either useful or interesting.

Insects are so intertwined with plants, soil and other animals in a habitat that watching insects leads us to the entire gamut of life in nature. So, watching insects takes a person very close to nature. This book is written with as minimum jargon as possible, to introduce readers to the world of insects especially students and lay public. However, to avoid certain technical terms is difficult and therefore this may pop up here and there. But a single reading through the book will certainly show how varied insects are and this variation by itself should be a stimulation to go out into nature and watch these creatures. A simple hand lens, pen and pad are all that one requires to watch and record insects.

The web offers myriads of sites where one can further access insects' knowledge. The website www.nbair.res.in is one place where more than a thousand pictures of Indian insects are featured. The pictures in this book are drawn from there. Readers are encouraged to also browse the website of Amateur Entomologists' Society (www.amentsoc.org). Here search for identification service and basic field keys which are very useful for beginners. Through this book, the author expects to enlist as many amateur entomologists as possible in India and elsewhere to further the cause of insect study and conservation. The book, though with Indian examples, is equally relevant to all novices around the world who wish to watch insects as a hobby or pastime.

I am happy to acknowledge the valuable insights given to me by several teachers who taught me entomology including stalwarts whom I've never met but their books during my student days were very informative and inspiring. I have freely used the pictures of insects from the web site of ICAR-National Bureau of Agricultural Insect Resources (NBAIR). I also thank my doctoral students, Viyolla Pinto (who mostly wrote the chapters on social insects), G.T. Geetha, M.A. Rashmi and J. Cruz Antony for all the help while preparing the book. I thank Dr J. Poorani and team for most of the photographs in the book.

So, Happy Insect-Watching...

Abraham Verghese

2. Preamble

Insects are six-legged creatures and hence are called hexapods (*hexa* = six and *poda* = legs). They are everywhere, yet most are less noticed except the ones that nag us - the cockroaches, mosquitoes, houseflies etc. As a birdwatcher I know, a bird to insect ratio is one to a lakh or more. Yet, birds are more watched. The logic is probably that the feathered bipeds, as birds are known, are colourful and vivacious with sweet calls and interesting behavioural patterns that attract attention. These however also hold true for insects. Yet, they are not watched except mostly by professional entomologists for academic reasons, and less for the sheer joy of watching them. It is pertinent to say that if we fail to watch insects, we miss a big chunk of nature-watching.

However across the world, a few groups of insects have drawn attention of enthusiasts beyond the perimeters of academia and insects of economic importance. These are the larger and attractive insects, mostly butterflies and beetles.

Insect farming, consisting mainly of butterfly parks not withstanding silkworm rearing and honeybee domestication, is becoming popular in many parts of the world, mainly for eco-tourism and insect-watching.

I have, as a boy always been enamoured by insects, especially ants. I enjoyed watching them trail along, quite reminiscent of a steam engine, forward bound, winding along serpentine courses. It, then never occurred to me why ants were not winged. But their locomotion was always an eye-catcher. I always enjoyed catching grasshoppers; their leap and jump sequence should envy any athletic coach training a high schooler to jump or pole-vault.

Cockroaches too have fascinated me. I always used to wonder (and still do) why they prefer crawling to flying when they have well developed wings. They too seemed to take to wings only to jump. Cockroaches were the one reason why study time late into the nights was sending creeps in us children in school days. No wonder I thought they were called creepy-crawlies. Our biology teacher used to say that cockroaches come from drains and sewages and I then knew why cockroaches were considered 'dirty'. I used to wonder why only drain, for sometimes they are found even below the mattresses. But what drew my attention then as a boy was the alacrity and alertness, these creatures displayed. Try to swat or stomp on one however stealthily, they invariably escape, leaving you exasperated and low.

I loved butterflies and used to catch them while at rest, or when sucking nectar from flowers with wings vertically folded. They are vulnerable. As a boy, I learned the art of catching them, mainly by watching the house lizards catching moths. The approach to the prey is slow but focussed on the target. There is a kind of hush-hush and on closing in there is a swift oral seizure. In a trice the moth is almost gulped by the vermin. I thought that was the way to catch butterflies. I knew they haunt flower beds of zinnia, marigold and some rose. Squatting alongside these shrubbies, I quickly learnt the art of catching the beautiful butterflies. They were not scary, but I used to wonder, how my holding their wings made their wings almost

Watching Insects

transparent. Little did I realise the coloured powder that dusted on my fingers were scales that dressed these creatures in various hues. Butterflies, are too "innocent" to be maimed, that I soon learnt how to hold and release them as gently as possible. If they flittered off, it meant I did'nt damage them.

Lady bird beetles were yet another attraction. They are a miniature look- alikes of the back-engined 'beetle' cars that were quite popular in the sixties of the twentieth century - well even now. These insects are coloured and spotted and slide slowly on grass blades and leaves. If held on ones palm they sure look lazy but, before one could feel them, they escape in a loopy flight.

My first encounter with a dragonfly was in our bedroom. One poor fellow found his way into the room and in a quest for escape was lured to the glass ventilator. Light obviously attracted them. But why not, I mused, the dragonfly never reverted to freedom through the door it entered, the doorway being equally lit and bright. But once at the ventilator, it kept whirring and flapping, and that's when it caught my attention. I tried to reach out for it in vain (as ventilators of those days were really high) but enjoyed watching it. It was whirring and striking against the ventilator glass between bouts of resting on its panes. The escape whirr seemed to taper off and the dragon fly seemed to give up, just resting as the evening sun dipped to shut the light, off the glass window panes. The next morning I found it dead. My parents told me it is the *"thumbi"* (The vernacular for dragonfly in Malayalam) and it would hold small pebbles on its "feet" if offered. I carefully picked up the fellow and found its wings coloured but transparent, body longish and eyes huge.

I soon found myself chasing these fellows till I felled one. Unlike butterflies their wings were stretched out on the sides. I held it by its wings and offered it a small pebble. It did "hold" just trying to use it as a landing but the same trick worked better on grasshoppers.

These childhood encounters were my interactions to the insect world. I loved them somehow and never realised as s school boy that I would eventually become a professional entomologist.

> Browse through Facts About Insects and Bugs - Funology - www.funology.com/facts about insects and bugs/; Fun Facts About Bugs- www.si.edu and www.goddidcreations.com for funny facts about insects like the following one:
>
> House flies find sugar with their feet, which are 10 million times more sensitive than human tongues!

3. Insects and Their Interesting Body Parts

Insects are creatures in the animal kingdom that have three distinct body parts- head, thorax and abdomen. They are the only creatures that have six legs (Hexapods) and two pairs of wings. They have a segmented body. Insects have jointed appendages and belong to the Phylum Arthropoda. Related to insects are the spiders and mites (with eight legs) and the many legged centipedes and millipedes etc. It is pertinent to mention that 2/3rd of the animal kingdom is constituted by insects. From the time of their hatching or "birth", insects keep changing for the better, till they become adults! These changes are called metamorphosis, a big word for the uninitiated. Yet this is so dominant in the insect world that one needs to know about it. Metamorphosis is nothing but shedding the external coat on them, loosely their external skin (also called the exoskeleton). These don't grow as insects grow. So, insects shed them as they grow along in 2-6 intervals. Between intervals, these stages are called the instars of larvae or nymphs. An instar of a larvae graduates into an adult but not before a resting stage called the pupae. Butterflies, moths, flies, bees, beetles etc have a pupal stage. There are some insects like bugs, aphids, cockroaches which have no pupae. Here the growing stages are called nymphs and invariably resemble the adults except that the nymphs have no wings or bear incomplete wings rendering them

Watching Insects

flightless. The exoskeleton is a unique feature of insects enabling their adaptation to a wide variety of climatic conditions.

Insects as adults are so numerous and varied. Equally varied are the larval and nymphal forms that these compound their variability to such great extent that no biota may display such a huge range of variation. So, there is almost an infinite array of life to be watched.

A species of the Papilio butterfly for example begins its life cycle on a citrus or curry leaf plant. The eggs are creamish and slightly smaller than mustard seeds. On hatching, ant like caterpillars emerge, which on metamorphosis and growth look like blackish bird droppings. Further as it grows, it develops into a smooth, green caterpillar and then pupates into a hanging chrysalis, dangling but firmly attached to the plant. From egg to pupa it must have passed into five instars. A few days after pupation, or the chrysalis formation, the pupa begins to show life. The butterfly formed in the pupa, struggles out and eventually emerges to freedom as it takes its first aerial borne adventure. The progression from an egg to a worm, to a winged wonder is certainly a magical transformation, wrought only by the Creator, for what else could explain the measly messy mass of a caterpillar, eventually alighting and flipping across flowers?

To appreciate this, one has to rear them. My first adventure into rearing these butterflies was when I reared the sulphur butterflies (Pieridae) in a cage. I remember Dr. Prof. G.P Channabasavanna, an entomologist of great repute, at the Entomology department of the University of Agricultural Sciences, Bangalore, who gave me a slot in the post graduate laboratory even while an undergraduate. He encouraged me to rear and observe the butterflies metamorphose from

Watching Insects

an egg to an adult.

These Pierids laid eggs on cassia shrubs that grew wild in the University campus. It was fun rearing it, as by then I was taught the rudiments of basic entomology. Rearing periods were really an eye opener. I watched it grow from an egg to adult almost by the hour. Metamorphosis and shedding their (exoskeleton) was so vividly etched in my mind, that even after three decades, insects as a unique creature of God continue to prop up in my mind prompting and sustaining my interest in entomology.

Another group of insect that impressed me was the weevils, especially the *Myllocerus*. The adult feeds on the leaves and the larvae (grubs) are mostly within the soil or plants. Whenever I attempted catching an adult, they characteristically feign death, a very interesting behaviour. The weevil just freezes and like a naughty boy ducking into hiding when caught in an act of mischief, these weevils just drop on to the grass below. Nine out of ten times, it is difficult to retrieve them or search amongst the grass or clods of soil. They remain motionless till probably I leave! Yes, this was their way of escaping from predators.

Insects and birds have two things in common. One is flight and the other sound. Generally calls and songs of birds are deemed far sweeter, hence the reference to "nightingale" like voice. Calls of the koel and robins, chirping of sparrows and even quacking of ducks all sound pleasant. But the sounds produced by insects either go unheard or are ominously deemed creepy. However, there are quite some musicians in the insect world. The stridulations of crickets, for example, from dusk to dawn is creepy to some and music to

Ash Weevil: *Myllocerus undecimpustulatus*

others. The continuous screeching of the cicadas, heralds the advent of summer or winter. Interestingly, in Bangalore, it coincides with the flowering of the Mayflower (*Delonix regia*), (but the two are not related!) heralding the arrival of summer.

Cicadas are large bugs and as adults have a shorter life span. They seem to spend all their adult life screeching, each screech lasting for long spells. Interestingly, cicadas are a delicacy in Thailand. I once saw a lady in Thailand chasing cicadas with a crude hand net. On enquiry, I learnt, these are caught, dewinged and eaten as fried crunches. Talking of Cicadas as food, I would venture to say that there are nearly 2000 species of different insects that are eaten as food. In India too, insects are eaten but I saw the real insect eating on the streets of Pattaya, Bangkok, where fried grasshoppers and scorpions were sold literally on the streets.

Globally, the most commonly consumed insects are the beetles (Coleoptera- 31%), caterpillars (Lepidopters- 18%) and wasps, bees and ants (Hymenopters- 14%). Following these are the locusts and the crickets (Orthoptera-13%), cicadas, leaf hoppers, plant hoppers, scale insects and true bugs (Hemiptera- 10%), termites (Isoptera-3%), dragonflies (Odonata- 3%), flies (Diptera-2%) and other orders (5%). Insects are highly nutritious and healthy food source with high fat, protein, vitamin, fiber and mineral content. The composition of unsaturated omega-3 and six fatty acids in mealworms is comparable with that in fish (and higher than in cattle and pigs) and the protein, vitamin and mineral content of mealworms is similar to that in fish and meat. (For more details on insects as food, please refer to- Huis, A.V., 2013, Potential of insects as food and feed in assuring food security. *Annual Review of Entomology.*, **58**(1): 563-583)

The house fly, (*Musca domestica*) is abhorred by all for its nuisance. If you get to see its young forms called maggots in rotting debris, you would only to tend to be repulsed. However, these are increasingly becoming popular as animal feed, especially in poultry, livestock and fish. Insects and insect-derived products have been used as medicinal resources by human cultures in many parts of the world since ancient times. Science has proven the existence of immunological, analgesic, antibacterial, diuretic, anaesthetic and anti-rheumatic properties in the bodies of insects. Several authors have studied the therapeutic potential of insects, either by employing insects and their products at the laboratory and/or clinical levels. Thus, insects seem to constitute an almost inexhaustible source of pharmacological research.

Insects have a striking presence on earth both in terms of diversity and abundance. Little do we realise this, for the mosquitoes, cockroaches, fleas, and houseflies dominate our attention, with perhaps an exception of honeybees and butterflies. There are many who don't realise that silk is from the saliva of an insect. Then there are some who think spider is an insect, which is not! The first step therefore, in watching insects is to know what an insect is.

Insects are the only creatures on this planet which have three pairs of legs, two pairs of wings and one pair of antennae as mentioned earlier. However, exceptions only add to the intrigue and fun for an insect watcher. The flies as a group are called dipterans (di=two; ptera=wings). These also defy the definition as they have only one pair of wings. This should excite us! If a fly is looked at carefully, a pair of knob like structures appears from the sides of the thorax behind the forewing. This is the highly reduced hind wing and entomologists call it halteres. This in fact helps in flight as a balancing structure.

The huge variation in the type of legs, wings and antennae only add to the fun for an insect-watcher. Remember, the body is divided into three parts- head, thorax (the middle part) and an abdomen. This is characteristic of all insects. Insects have no skeleton unlike birds, reptiles and mammals that include humans. Their shape and size are governed by a hardened chitin structure outside the body. So it is called the exoskeleton. Some insects have an elaborate social system. There is a queen, king (males), workers and soldiers. Examples are ants, bees and termites. These invariably live in colonies and would have a single queen. The queen 'governs' the colony proceedings - truly a matriarchal system. The 'king' mates and dies, for he is indeed 'useless' to the colony after that! The workers have to gather food and defend the colony and are in most cases (like the ants and termites for example) wingless defying the definition of an insect but the three pairs of legs redeem their status to the insect class.

However bees are an exception, as workers also have two pairs of wings. Apart from different species of insects, the diversity in forms, colour, modifications of legs, antennae and wings and other body parts make watching insects an exciting and absorbing pastime.

> Wasps feeding on fermenting juice have been known to get "drunk" and pass out!
>
> Insects are cold-blooded animals. To survive extreme colds, some insects replace their body water with glycerol which acts as an "anti-freeze".
>
> 1.5 million insects have been named. Of these around 60,000 occur in India and 91,000 in USA.
>
> (www.si.edu)

4. Insect Antennae

Insects have a pair of antennae on the head, that typically looks like a pair of horn on a cow or a bull. The antennae are sensory organs of an insect. Just as we feel heat or cold or the presence or absence of an object by touch, the insects too "feel" their surroundings with the antennae. During an encounter with an insect like say a cockroach, you might have come across a pair of antennae from the front-most region of the insect head. The antennae is actually a collection of highly complicated sensory network system that has the capability to orient or disorient the insect. In fact a lot of behaviour of the insect depends on how the antenna responds to the environmental stimuli.

Technically speaking, the antenna is made up of three basic segments. The scape (base), the pedicel (stem) and the flagellum. The base areas of this very antenna we are talking about is the area where it attaches itself to the head of the insect. Like the foundation of the building is entrenched into the earth, the base of the antennae slots into the exoskeleton of the head. The pedicel or the stem is rightly like the stem of the tree, above the roots and holding the upper part of the system. The most dynamic and tactile region of the antennae is the flagellum, which might appear thick or tapering or with a club or is like a whip. The flagellum is the most sensitive region of an antenna and the area that bears most contact with the ecosystem or surrounding. Almost everything the insect comes across is gently touched or sensed by the antennae and the signals and messages are received. The function is almost akin to the signal receiver of a television which is why it is also called the antenna. The TV antennae of the yester years have given way to dish antennae. Likewise, insect antennae also have a lot of modifications worth observing. The flagella are a collection of smaller flagellomeres, the number of which is variant in various insects.

As you involve in watching insects, you might realise what significance the flagella has in grouping insects into their relevant factions. Every twist and turn of an antenna is imperative in Entomology. A bristle-like antennae seen in the dragon flies is called the setaceous antennae. It is the broadest at the base and it tapers out. The next similar sort of antennae is the thread like filiform antennae. The flagellomeres of this form of antennae too tapers out from the base but do not give a distinct bristle like or whip like look. The units of the flagellae are distinctly seen and can be observed in ground beetles and cockroaches.

Filiform Antennae in Plant Bug

The termites on the other hand have a collection of stubby-bead like flagellomeres arranged one on top of the other. Almost all are equally sized and this monoliform antenna decorates the head of some of the beetles. They are usually shorter unlike the filiformed ones.

A saw-tooth arrangement of flagellomeres is seen in click beetles. Serrated appearance is because of the arrangement of the uneven shaped flagellomeres one on top of the other. The carrion beetles have a reverse monoliforms sort of antennae. It looks as if the bigger beads are away from the head poised elegantly over the smaller ones. This arrangement is called the clavate form.

Clavate Antennae in Blister Beetle

The butterflies bear the capitate antennae. It is abruptly clubbed and broader units are seen on the distal tip. Ants have a distinct "elbow" shaped antennae. The scape is distinctly longer and connects the units and the head.

The pectinate antennae of a fire beetle, resembles bristles of a comb which are slightly curved and with radiating bristles on one side of the antennae. The plumose antennae of a male mosquito are dense and feather-like and to some it gives a "cottony" appearance

Elbow-shaped Antennae in Weevil

at first look like a "feather on an English hat". The antenna of a house fly bears a lateral bristle like extension on a small baggy pouch like antennae. This helps them to smell their way into our eateries, kitchens and dining table!

Pectinate Antennae in Silkmoth *Attacus atlas*

Watching Insects

For the size of the insects it should go unsaid that you will definitely need a microscope for a joyous view of what an antennae looks like, but a good hand lens is sufficient for most of the bigger ones. As much as insects are varied, so are insect antennae, adding to the joy in watching them.

About 70% of animals are insects!

Only about one percent of all insects is harmful.

They say a cockroach can live upto three weeks without its head!

Some ants can carry 50 times its body weight!

Ants don't sleep!

Flies have 4000 lenses in each eye!

5. Insect Legs

"Everything that has the capacity to get a screech from me is an insect", a student had once told me. I am very sure she included almost every creepy crawly (meaning insect) in her list and may be even spiders too which are actually not insects.

Well let me remind you once again, only the six-legged creatures are insects. The three pairs of legs an insect bears on the middle part of the body are all jointed. The thorax (mid part of the insect body) is divided into three segments and each segment bears a pairs of legs. The phylum Arthropoda to which insects also belong means (in Latin) jointed legs. The other non-insect arthropods are the spiders, ticks, scorpions, mites, millipedes, centipedes etc.

Considering the capacity of insects to have flourished in almost all parts of the earth, I must say their locomotory organs have a major contribution. Insect legs are highly modified for different functions, depending on the environment and habits of the insects. Observing the legs alone is fun and insect-watchers do well to know a little more about their legs. The legs or the appendages as they are also called are made up of five segments called "tarsomeres". The three pairs of legs arise from the fore, the mid and the hind regions of the thorax. The main parts of the insect legs are the coxa, the trochanter, the femur, the tibia, and the tarsus. A little more for them who would like to delve into the biology of an insect is that the coxa is the region where the leg and the body are linked.

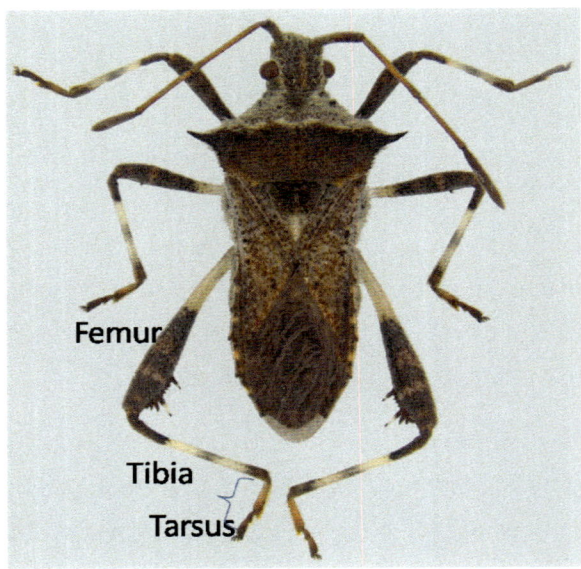

The diversity and widespread distribution of insects are supported by the modifications one can observe in locomotion. A few swim, a few jump, a few leap, a few crawl while a few can float on water. May it be a flat smooth surface or a razor sharp edge, an insect can ward off any intricate condition with appropriate manipulation and co-ordinated movements of the five parts of the leg. Yet there are a few insects like the scale insects which in their completely leg-less condition are called sessile. Some in the larval forms, for example, "worms"-(actually maggots) in fruits are also legless!

Fast running insects like many bugs and beetles bear ambulatory legs. The femur and tibia are strong and they not only aid in movement of the insect but slightly lift it and propel it forward to quicken the pace. Watch closely a beetle or a bug or even a cockroach and you will be surprised as it vanishes at the wink of your eye.

Now a little more on the all too familiar cockroach. All your attempts of killing the pesky roaches in the kitchen would have gotten on your nerves. Just to make you realise that your muscular movements cannot out-beat that of the roach. Thanks to its cursorial legs that make these swift runners. Long thin segments help them sprint to safety. The next time you are in your kitchen with a swat or a broom, you know what should be running in your mind. For those who poison the roaches in the kitchen, remember that the toxin affects the rhythm in the locomotion leading to its death. Well and if you miss swatting a fellow, humbly accept its superiority in fleeing from danger (or you!).

Fossorial legs have evolved for example in the ground dwelling bugs. The mole crickets and the younger cicadas are highly equipped for digging into the soil. Their legs are broad flat and dense. Soil is dug out and the insect shelter in small caved soil burrows.

Natorial legs are for the swimmer. A broad oar is made up of long strong setae on the tarsi. Aquatic insects like the aquatic bugs and beetles are equipped with these for movement in water. Ever seen a giant water bug? They bear oar-like flattened legs.

To hunt, an insect has to have appendages where the insect should be able to clasp. Rightly so, preying mantids have fore legs modified to grasp prey. Now most often people mistake it to be a "praying" mantis; for it seems that the insect is praying or pleading, from a lateral view. Well maybe so! These modified legs are the raptorial legs. The clasp would be quite tight to let go the poor prey, often an innocent bug or butterfly that lands near an opportunistic mantid.

The saltatory leg of a grasshopper is the envy of every athlete. The strong femur on the hind leg is distinctly visible and the femoral muscles sure account for the jumps seen in grasshoppers, crickets, katydids etc.

The other type of modification are claws in the biting louses. In bees corbiculate legs are seen where a special expansion forms a pollen basket into which bees tuck in pollen from flowers and carry it back to the hive to feed its young ones. Pollen comb and a heavy pincers are seen on the lower tarsomere which cleans out any pollen stuck on the legs.

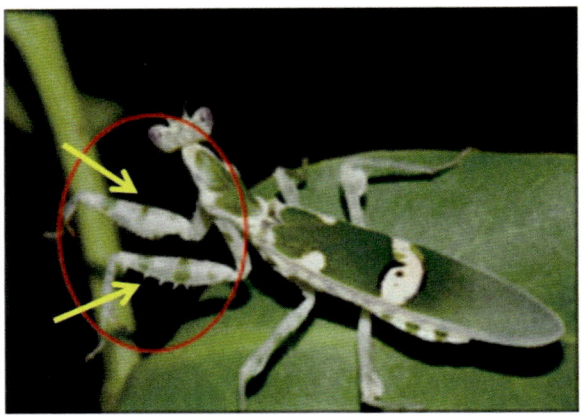

Raptorial Legs in Preying Mantids

Beyond locomotion, the legs are used for making sounds for interspecific communications, especially in courtship. Strong kicks are given by some hoppers, stink bugs and the tree hoppers as a defence mechanism.

The legs bear spines and a large grasshopper can cause you a little pain if you try catching it with your bare hands. The insects groom themselves with the help of their legs. Cuticular combs and setae extensions help in keeping the insects busy in grooming in their leisure. Most of the social insects indulge in grooming.

> **Butterflies taste with their hind legs**
> www.goddidcreations.com

6. How to Watch Insects

Many ask the question "how to watch insects?" Most people don't or will not! If an insect, whatever species scamper along, the first response is to shoo it away or run (for life!) but never to stand still and watch it. Or, the response is to get a swat or fold a newspaper and swat it and kill it. (How cruel!) Harmless crickets, bugs, dragonflies and moths, meet this end. Even many of the advocates of ahimsa make insects an exception. They kill them invariably. However, if we learn to watch them, many prejudices against them vanish. But the big question is how to watch insects. I take this on a group by group basis.

Equip yourself with a hand lens 10x or more (with or without light), a note book, a camera (your mobile camera comes handy) and a pen. Now get ready to watch!

7. Crawlers and Earth-Bound Insects

There are many insects that spend almost all their lives or most of it on the ground. Some of them have strong legs and move around. Such of these are the ants, ground beetles, earwigs, termites, cutworms, etc. The ideal place would be to select an undisturbed patch off-track of a path in a jungle or right in the middle of a meadow. Here just squat and wait! It is quite possible that a ground beetle or a scout ant will just appear sensing for food. Insects usually don't move without cause. It is always in search of food, prey, nesting site etc. Squat and wait bring insects to you. What is important is to take note of their description as soon as possible, their food, their habits, behaviour etc. Now-a-days, it is possible to even follow or video these insects even using a low end mobile phone. These can be compared with pictures in this book or later consulted with experts or even browse insect websites. Most of the common insects can be found on the National Bureau of Agricultural Insect Resources website (www.nbair.res.in)

8. Insects of Grasses and Herbs

Many insects may avoid open sites, clear areas except to cross-over. Quite a large number of species are partial to grasses and herbs. Here again, squat and wait helps. A sharp eye may be able to spot a stick insect, for instance, or a pair of ladybirds mating or a trail of ants or a bee pollinating or even a butterfly flitting from flower to flower of small herbs. You may be surprised to see that a few square metres around you is beaming with millions of life. This can even prompt you to have a mixed patchy land with diverse plants in your backyard. The patches hitherto considered weeds and hence to be removed, becomes a reservoir of the Creator's Noah's Ark out in open! With more types of grass and herbal plants in low patches you are bound to see more species of different insects. With more diversity in plant life, there is bound to be more diversity in insect life! Now do you need a well manicured lawn or a herbal patch of diverse flora? The latter hardly needs maintenance yet gives you more hours of insect-watching. It is better to avoid the use of insecticides in and around the gardens. This will enhance more insect diversity. More the insect diversity, less the number of insect pests!

9. Watching Nocturnal Insects

The patch you spend watching insects for hours a day might offer a different array of insects once dusk sets in. You need a single light source fixed in the dark or, a torch, a good pair of boots on for protection against equally interesting nocturnal vermins. Below street lights, widely spaced out in villages or a garden light (with lights inside houses put off for some time) also serves to attract bugs, moths, beetles etc. It is easy to watch many nocturnal insects photo-orient to light sources and keep flying or hovering for hours. They may even drop dead in fatigue, as for some, light gives a heady and strong feeling that they keep beating around it! Even a city dweller will wonder how many types of insects, of varied kinds occur in concrete jungles or in a fully urbanised city. So a light source during night is a rallying point where hours can be spent watching flying nocturnal insects.

10. Watching Insects Visiting Flowers

Flowers and insects invariably go together. Insects like bees, flies, butterflies, ants and even wasps visit flowers for nectar and/or pollen. So an undisturbed bed of ornamentals or a patch of flowers attract a lot of insects. There is no time of the day or night to watch insects at flowers. They come in peaks and surges and then harmonically lowers to again peak. It is good to find a spot and squat or follow a trail in a garden. Unlike birds, insects are not easily disturbed, nevertheless your pace must be slow. There are a whole lot of plants that can attract insects for you to watch. Have a few seeds of sunflower sown and once the flowers open an array of insects come to it including the nocturnal ones. Generally, there is no competition among the flower-visiting insects as they stagger on a time frame. But bigger bees or insects can displace smaller ones physically. An insect at a flower is barely territorial as each insect is too concerned of its food. Hence they will not try to drain away energy to chase the other out, as there is abundant supply. So, one can have a diversity of insects on a big flower, like the sunflower. There are many other plants which can be planted to attract insects and some of them are, the flowers of cassia, calendula, poppies, cornflowers, the common fennels, roses, sunflowers, the blue cluster vines, *Plumeria*, the *Delonix* - May flowers, common *Jacaranda*, *Pongamia*, etc. It is no exaggeration to say that all plants attract insects. Some more and

Watching Insects

some less.

Try planting some of these and as they flower, insect-watching becomes a rewarding experience. An insect hovering above the flowers and settling for a little nip of nectar is the time one can get a close look at it. As the little fellow lands on the flower, you can invert a transparent vial over it to trap it for a moment and after a closer look release it back to its foraging mission. Take care not to damage it. Use a bigger tube for bigger insects, lest you damage the fellow. This might be easier if you spot an ant foraging on a flower. Since these are not easily disturbed, you may stand by the plant and quietly watch the little fellow walk to various parts of the flower, pause, gets back, re-orients itself, grooms itself free from the little pollen sticking on to its antenna and legs and then decide to scurry down the plant. One advantage with the ants is that since the usual workers or foragers are wingless, they can be followed through the trail it takes. In one journey the little one may visit more that 50 flowers and then get back to its nest. Following it may lead you to the ant-nest that may be close to the proximity of the flower or way further. On finding food more than what the lone forager can take back to its nest, it may even travel back all the way to its nest leaving a trail of scent to get back to the same spot with a few more nest-mates following it to carry the finding (food) home.

Watching ants can consume your time to such an extent that you would be surprised of not realising the hours that passed by. The sight of the ants caressing the little aphids that excrete out little droplets of sugary substance is worth watching. The fragile movements of the antennae moved on the backs of the aphids, the gentle taps of soothing movement on the aphids show the intricate give and take relationships these insects have. While offering protection to the aphids against predators, these ants fill their bellies with the sugary water and retreat to their nests.

Crotalaria Attracting Butterflies

11. Watching Flying Insects

Insects like birds have wings. In fact, insects have one pair more! Birds with one pair of wings fly more and migrate between continents. Insects too migrate and sometimes beyond geographical frontiers. Insects in their younger stages as nymphs or larvae are without the capability of flight. In insects, nymphal stages at the best may have a pair of wing buds or pads till they metamorphose to adults. For an insect watcher, following insects in flight is quite an absorbing pass-time. Watch a flying dragonfly. It is best to start from the time you spot a resting dragonfly, may be on the boulder usually near water sources. You squat beside and soon the dragon fly takes off and is back on the boulder! It will repeat this art again and again. You may wonder why? I too do. But it is this sense of wonder that keeps you engaged in insect watching. It may be a courtship tactic or a move to stalk an unwary flying but smaller insect, as dragonflies are predators. It is not uncommon to see a pair mating on the wing. Equally absorbing is the damselfly.

Many insects, especially the butterflies migrate. Swarms can be seen in irregular flaps in criss-cross flights but oriented to a definite direction. They are not goal-less. It helps to use a pair of binoculars especially the ones with wider field. 7 x 35+ is recommended. (This is the same as bird-watchers use)

Sometimes in field you may encounter swarms of rock bees, as these two move in search of new nesting sites. Don't panic is the *mantra*! If you are caught in their flight path, be still till they pass over, lest you get stung. But appreciate the fact that these bees pollinate many flora, enhancing yields of many crops; otherwise we would have been without fruits and forests.

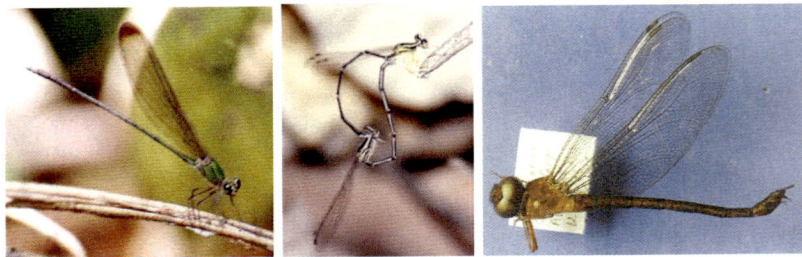

       Damselfly   Damselfly Mating   Dragonfly
(Courtesy: Dr. K. Veenakumari, NBAIR)

Dragonfly has a pair of big eyes while a damselfly has a pair of smaller eyes.

12. Collection and Preservation

While in the field, a small glass jar (=500 ml) or a glass tube of at least 10 centimetres with a wide mouth (>2.5 cm) and a stopper/cork is useful to trap small insects and observe them using a hand lens to note their description or even photographing them. Sometimes insects under soil debris or clods need to be raked out gently. By inverting an open glass jar over them, insects like earwigs, staphylinid beetles, ground roaches can be temporarily trapped and observed. To observe smaller insects like thrips inside flowers is difficult; some gentle taps on the flowers with our palm below, will dislodge them for us to watch. All insects caught should be released back to their original habitats without injuring them.

Insect-catching, meaning collecting, pinning and labelling them and then storing them in glass topped wooden boxes have been both a professional and amateur pursuit. However, I strongly disapprove of this for an amateur, for whom this book is written. Insect watching is fun, less clumsy, more nature-oriented and healthy. These days even an ordinary mobile phone captures an insect picture, especially the larger ones, which is almost sufficient to identify them, if of course, they are known, for the probability of getting a new insect is very high, as we are still to know and name nearly 70,000 insects plus! However, I encourage collecting dead insects, especially at light sources and out in the gardens and fields where spent adults die. If these insects are fresh, they can be preserved. Steps in preserving dead insects (not killed):

- Wash with a brush and alcohol to remove dirt, dust and soil particles, fungus and other adhering debris.

- If dry or brittle- relax them, pin and dry. Relaxing can be achieved by shutting the insect in an air-dry wide-mouthed, bottle with a wad of wet cotton placed to the sides. Care should be taken to avoid contact of the insect with wet cotton. This prevents fungus formation on the dead insects. Normally in a couple of days, insects are relaxed which can be made out by less brittleness and more pliability.

- Stretch them on a stretch-board or on styroform, as symmetrical as possible, and pin them. A sewing needles with a bead stuck on top serves as an excellent insect pin. Common pins rust and are hence are not suitable.

- Label them using the smallest fonts or use black Indian ink. Labels should have on them the place of collection, type of habitat, name of the collector and date. Another label could be of the insect identified. The labels should be of the size less than half square inch

- Pinned insects should be kept in boxes with a soft base (cork board styrofoam, etc) These boxes should be preserved in an air-tight cupboard. Transparent plastic boxes with styrofoam base also serves the purpose.

Watching Insects

You can make insect boxes, in which small pouches with powdered naphthalene balls or camphor should be kept to keep away fungus, mites and other microorganisms.

To know more on insect mounting, stretching and pinning some useful websites may be accessed through search engines available on the internet. I have found these to be quite informative.

Excited Students Watching Collected and Preserved Insects

13. Making of A Good Insect Box

A plastic bread box (8"x6"x4") with a soft base like cork board with a transparent lid serves as an insect box. If the top is transparent, it helps to see the insect even without opening the lid. For all that interest that's begun to build up in you, at least a few of you might be thinking of making a good insect box. Let me now introduce you to making a small insect collect box for yourself which enables long term storage.

14. Insect Box for Storing Insect Specimens

You have two options: A neat, shallow cardboard box will work, or you can use a glass-covered wooden display case taking help from your local carpenter. Its nice to have a soft styrofoam or strawboard base as mentioned earlier. To pin an insect in place, firmly poke the pin through the upper mid-right portion on the dorsal side of the thorax (on insects such as grasshoppers) or upper side of the right wing (on a beetle). Use tweezers or forceps to handle small specimens.

Use a dab of clear glue to stick really small insects onto the apex of a triangular pointed card (A triangle with 15 mm length and 7mm base) and then pin it with the dorsal side facing up the card near the base. Specimens within a group may be neatly grouped above the group label or assorted size wise. Pointed specimens face to the viewer's left, but their labels are parallel to the labels of the pinned specimens. Although no specific size of the label is mandated for an amateur, most attractive labels are approximately less than half an inch long and 5/16 inch wide. Pin the labels directly a little above to the floor of the box and below the insect, and arrange all specimens representing a group in neat rows in a rectangular manner. (Always arrange the insects across the length of the box).

I strongly encourage visit to insect museums where dead insects are stored to get an idea of insect curation and storage. Almost all agricultural colleges have an entomology wing with at least a small museum.

Insect Box with Pinned Insects

15. Watch Grasshoppers and Crickets

One of the best places to watch grasshoppers and crickets are the grasslands and meadows. As you walk through the grass your feet flush out myriads of short-horned grasshoppers, long-horned grasshoppers and crickets. To this group belong the dreaded locusts. Though they move about jumping and by short hauled flights, locusts are capable of long range movements, even between countries.

One of the characteristic features of a grasshopper is the "clicking" sound it produces as it jumps and takes to short hauls of flight. During this jump-fly locomotion, the edges of two wings rub against each other to produce the sounds. At rest they also produce a sound by stridulating the hind legs; this keeps the courtship going in a pair. Isn't it wonderful to know that grasshoppers can hear? Somewhere on the abdomen they have a pair of "ears"- the auditory organ!

One noticeable feature is the well developed femur of the hind leg. This third pair of leg is a miniature form of an ostrich leg and enables the grasshopper to leap high forward. Invariably, the thorax is well developed with a shield-like look and rests on the grasshopper as a warrior's mantle. The upper or fore wings are 'leathery' while the hind wings are membranous. If you ever get to handle a grasshopper, gently flip the upper wing, and below is the hind wing which at rest is worn like a draped and pleated frock under the upper pair of wings.

While in flight, these are outspread. Obviously the lower wings are larger. These can be seen only when the insect leaps off in flight.

The most common grasshoppers are the short antennae (horn) grasshoppers (also called Acridids). These are widespread mainly in the tropics and subtropics and frequent grasslands, low bushes and may be found feeding on low tree canopies. They are also found on dry deserts, rocky terrain etc. They are quite common in grazing pastures and cattles flush hordes of them out. In India, for this reason, the bird, cattle egret (*Bubulcus ibis*) follows cows and buffaloes as they graze, to pounce on an unwary grasshopper that is flushed out by the grazing.

Shorthorned Grasshopper: *Acrida exaltata*

The upper or fore wings are green to brown coloured, ideally to render them obliterated in their surroundings. The other pair of wing is brightly coloured as though to compensate for the dull colour of the fore wings. The young ones of the grasshoppers are wingless and as they metamorphose, wing pads develop to finally moult into a full winged adult. Many times, the young ones or the nymphs may have colouration totally in variation to the adult. So, the best way to identify a species is to observe a fully developed grasshopper. But watching the nymphs which are smaller but resemble the adults, have equally strong legs to jump, is fun. Yet there are adults which are wingless like the Deccan wingless grasshopper *Colemania sphenaroides*. In some adults, the wings are merely half developed. (Ex: *Hieroglyphus* sp).

The grasshoppers including locusts are prolific breeders. The females lay clusters of eggs in a hard bag like structure called ootheca in the soil. So, these are reasonably more protected. On hatching, the tiny nymphs of grasshoppers, (which typical of a nymph are wingless) start feeding on the green plant material being totally "vegetarian". They pass through 4-10 months as nymphs depending on the species, to finally become winged adults.

The grasshoppers being green feeders, become pests on crop-plants. However the abundant meadow lands, in and around of agro-ecosystems will enable them to survive the anthropogenic pressures like ploughing, harrowing (which destroys the eggs), pesticide pressures etc.

The long-antennae (horned) grasshoppers essentially differ from the Acridids in that it has a pair of antennae that is as long as or longer than the body length! The upper wings slightly slope to the sides! These too are found in habitats frequented by the Acridids, but the 'songs' are generally less clicky during flight and while stridulating produce a sound, as the English have described it to be *Katy did*! Hence these have been also called Katydids. Most of the long antennae grasshoppers are beyond 4-5 cm in length. These grasshoppers are less sub-terrain. They lay eggs on plants and grasses, barks of trees or plain surfaces of soil. Some species are tree dwellers. These being plant-feeders are victims of pesticides as non- target insects. Small patches of undisturbed meadows will help conserve them as not all grasshoppers are pests.

Longhorned Grasshopper: *Phaneroptera gracilis*

These as primary herbivores (including Acridids) form important food links in the food web. Many reptiles, amphibians, birds and small animals depend on grasshoppers for food. Though these can be sighted during the day, most of them are active in the night, and can be watched with a torch flashed out in the dark, by merely flushing them out by walking through meadows and grasslands. As they become more active during the wet weather, walking with a pair of boots helps. It helps to have a stick in the nights especially to off lodge an unwary serpent or some night vermins, which even while on approach normally slither or scamper away. During the nights, along with the crickets, their "*Katy-did*" songs in different octaves and notes ring out, but familiarity and expert help are needed to properly identify them. Even if not sighted, they add to the 'sounds of the night', mostly with a lullabying effect, especially for nature lovers. The 'music' or song of the grasshoppers helps in courtship between sexes.

> Swarms of locusts that move in African and Asian deserts can be in several thousand millions!
>
> They can cover at least 2000 miles in a season, and consume its own weight of food in a day.

16. Watch Crickets

Crickets (and not to be confused with the game cricket- singular) are found all over India and unlike grasshoppers, are more down to earth, as most of them live under fallen trees, stones or mud clods. Some live in burrows below. They are invariably dark brownish and merge within the earthly habitat. They may be found around human dwellings, houses and damp places. Crickets do resemble grasshoppers with long antennae, but their habitats of frequenting dry places and nocturnal habits give them away.

On examination, very interestingly, one can see the hindwings extending beyond the abdomen. The hind femur is developed but may not appear as prominent as in the grasshoppers. However, some are wingless even as adults. Crickets are good musicians and produce interesting sounds and several males of the species create quite a cacophony from late evenings into the night-(till you doze off!) while they are in a world of love of their own.

The noise you hear from crickets is called chirping. It is the sound of males trying to attract females. A female cricket may fly over great distances as she homes in on the calling song of a male. The journey could be risky, because it exposes her to predators, but it is necessary if she is to mate and lay eggs to produce the next generation of crickets. The noise is made when one of the wings of the insect is raised to an angle of 45^0 and is rubbed against the other wing. Now you will have to watch a cricket to understand the strumming of the music, but it is not easy, for on approach, by a human, the music is suspended!

They are omnivores and hence like cockroaches can be found as household pests. Some are pests on crops, and most of them live for several months.

There is yet another form of cricket- the mole crickets. These have a well modified foreleg easily discernible and distinguishes it from the common crickets. This pair of legs helps the insect dig and burrow into the soil. The habits are all like the crickets.

As I have mentioned, crickets can be seen around homesteads. One common species is the *Gryllodes* and the other *Trigonidium*. *Gryllodes sigillatus* is a spotty, dull-yellow to brown colour cricket. The females lack wings. At the tip of the abdomen there are a pair of tail like processes easily discernable. The females look three-tailed because of a long ovipositor in the middle. These too like grasshoppers jump. Short crawling and then a leap is characteristic. Sometimes they jump without crawling.

Cricket: *Trigonidium humbertianum*

A near relative, *Achata domestica* is more in the gardens and here both sexes have wings. The best time to watch them is in the night or at low lights or in a torch light in gardens. If doors or walls of bathrooms have cracks or holes, crickets will lay eggs in them. Literally, they breed where you bathe! Homestead crickets love to eat anything, being polyphagous. However, these are common in rural houses as modern apartments, sealed with air-conditioners are not habitable by crickets. But beware, they do adapt fast!

17. Watch Stick Insect and Leaf Insects

One of the queer insects in the insect world is the stick insect. These are herbivores and never gregarious. Hence they are not damagers of crops. They either look like stubs of dry grass or long shoots of meadow grass. It has a cousin, the leaf insect. While one looks like a stick with branches or dried grass with nodal stumps, the leaf insects are generally greenish with a leaf-like look with even the leaf-like venations. In both these insects, the mid portion of the thorax are greatly elongated and expanded. Most of them have very long antennae. Watching these is fun, if one can detect them, for often they merge with the habitat or the plants they frequent. This is an attractive museum display specimen. The large leaf insects from the north east *(Phyllium scythe)* is noted all over the world. Both the stick insects and the leaf insects are generally called the Phasmids.

Stick Insects   Leaf Insect
(Courtesy: Dr. K. Veenakumari, NBAIR)

18. Watch Earwigs

These small insects frequently sighted in soil cannot be missed by anyone. If winged, the hind wing fans out like a membranous semi circular "ear", hence probably the name. There are some who believe that these love to enter the ears of the people sleeping out in the fields. This however has no scientific basis. On careful examination, a Y shape is evident on the head, with well developed eyes. Some are not winged, because they do not need to fly, being mostly ground dwelling. To some the insect looks frightening and some even mistake it to be little scorpions! This is because of a forcep like process at the end of the abdomen, which the earwig uses to pierce or ward off predators. It is rarely used on humans, so without fear one can handle and watch an earwig.

Earwig: *Elaunon bipartitus*

Atypically, a female lays eggs and sits on it, much like an incubating hen and when the young ones hatch, they may still be below the mother's body, again typical of chicks under the wings of a mother hen. Earwigs are omnivores and get their food in the soil. In Malaysia, there is one group which is wingless and associated with bats.

19. Watch Roaches

Of all the insects, cockroaches are the most familiar. These dwell in houses, sewage and leaf litter in gardens. Most are commensal with man. Starting from school biology to higher sciences, cockroaches join the ranks of guinea pigs in taking forward and understanding of biology. They are teased with scalpels, cut with micro-scissors and anchored with pins and clasped with forceps to enhance biological knowledge! Cockroaches like an innocent lamb, led to be slaughtered, seem to bear it all. Yet, they are strongly disliked! These being strongly commensal, have survived all the onslaught by man, and so are to be commended! One of the reasons is their prolific breeding ability, as eggs are laid in a highly protected capsule, ensuring zero predation. Some roaches, even carry their egg capsules around till young ones emerge. Roaches are mostly active nocturnally giving us less room to notice or combat them. Many seem to have become resistant to insecticides. Most often, we the administrators of insecticides, suffer incognito. Residues of insecticides used against cockroaches remain in our kitchens and bathrooms and eventually enter in traces into our systems!

Cockroaches are so ubiquitous and familiar that one may frown on describing them. But these, considered primitive but a typical representative of Insecta do have definite characteristics. The antenna is filiform and long, with a visible pair of eyes. The first segment of the thorax is prominent and it overlaps the head. The lower wings are folded fan-like when at rest.

Cockroaches come in all sizes from less than a centimetre to large ones of more than four centimetres. There are many out in the gardens, in leaf litter and in grasslands (Example, *Therea* spp). These are active during the day and are considered useful as scavengers contributing to the nutrient cycle in the ecosystem. The domestic ones in our homes are adapted to a nocturnal life- they have to be or else they become easier victims of man's swatting (Example, *Periplanata* spp). Another adaptation is that these have easily taken to the food man eats; fruits, leftovers, bread, jam, jaggery, sugar, etc.

Honestly, watching cockroaches is interesting. They have long legs and are adept at running. Chase one, and you will be surprised at how they vanish in thin air but along the ground all on legs; rarely they fly. For all you know, they are just under your chair or bed. Many people inadvertently carry them to different places in their travel kits and baggages where cockroaches also hide.

Seven Spotted Ground Cockroach: *Therea petiveriana*

20. Watch Preying Mantids

This is another group worth watching. These can be easily confused for the stick insect, but characteristically have a well developed fore leg which is spiny, easily visible when at rest. While stalking a prey the fore leg is kept folded. Using these legs they quickly pincer down the prey and slowly enjoys the meal. It is an almost triangular headed creature which interestingly nods. Like the cockroach, the first part of the thorax is longest with a pair of long filiform antennae. Like the leaf and the stick insect these too merge with their immediate surroundings and plants. So some are stickish with twig like colours ranging from brownish to darker tan, while some resemble leaves, even early senesensing leaves with a dash of yellow on the wings. Some look like half wilted or dried leaves. For a hunting insect with other insects for food, this obliteration is absolutely an essential adaptation.

To see a preying mantis wait, watch and seize a prey is like a lesson in patience to a hungry man. The mantis rarely stalks but learns to wait for an unwary bee or fly to land near its outpost. Once the prey is sighted it is sized well with elegant deflexing of the head. Then positioning and gripping itself on the plant with the middle and hind legs, the fore leg is flexed slowly. The preparation is awesome, but crucial as the next prey might be hours away. To miss means to continue starving. From a vantage point, the mantis silently but deftly moves, accurately striking and clasping the prey between the powerful legs, which is lined with spiny processes. Once caught, the prey is chewed and swallowed. These can be commonly watched in gardens, woods and forests where they wait in camouflage like a soldier ready to ambush. One can easily approach a mantis unlike cockroaches, as they rarely scamper away. So watching them is easy as well as rewarding.

Praying Mantis

[Note: 'mantids' and 'mantis' and 'preying mantis' and 'praying mantis' are interchangeably used]

21. Watch Lice

Louses are indeed lousy as they are human and animal pests. So watching them means to watch out! The biting lice occur on birds, mammals and man and are essentially non-vegetarians! The bird louse is small and flat and live ectoparasitically on the birds feeding on feathers and superficial scaly matters on the bird. Sometimes they also feed on blood, probably from wounds. One species, *Haematomyzus elephantis* is found on elephants in the wild. There is another group of lice occurring on mammals including man. These have sucking mouth parts. These are flat and small sized and is commonly encountered in hair of humans. The head louse *Pediculus humanis*, is a scourge among poor people. These transmit the dreaded typhus by vectoring Rickettsia. It is also attributed to many types of fever. One species, *Haematopinus tuberculatus* is common on cattle especially buffaloes.

22. Watch Bugs

Bugs really don't bug except perhaps the bed bugs! This is an interesting group of insects, which feed mainly by sucking. The food therefore is always fluidish. If on man, like the bed bugs, the blood is sucked. On plants like the aphids, the plant sap is sucked. In many bugs (Heteropterans) one half of the upper wing is membranous. Some bugs are smelly as they release some chemical from a specialised stink gland. These are to repel the predators. There are quite some bugs, like the mealy bugs and scales which have mealy or wax "secretions" covering them. Often they are sessile (rudimentary legs) and hence hardly move. Erroneously, insects in general are called bugs by the lay public. Bugs are a definite group quite distinct from say, the beetles, lice, wasps, bees and so on.

23. Watch Assasin Bugs

As the name suggests, these assassinate! They wait in ambush, stalk juicy insects, strike them from behind or sides. Their main weapon is their mouth. The mouth is modified into a powerful pierce and suck organ. These even attack bigger animals and sometimes even man (but don't be scared, there is no harm done!). Some of these also called Reduviids are medically significant as they vector a parasitic protozoan *Trypanosomes*. African trypanosomiasis and Chargas diseases (sleeping sickness) cause several deaths in Africa and Latin America respectively. The species *Rhodnius* and *Triatoma* are well known, therefore in the medical world. However, there is no need to muse apprehension of the assassin bugs. They are wonderful creatures to watch and unlike the almost similar looking bugs, they do not strike! If handled, the reduviids can inflict a painful sting. People dwelling around woods and forests have reported these sucking blood in the night. However vast majority are useful as predators on other insects in agroecosystems.

Reduvid Bug: *Acanthaspis* sp

24. Watch Coreids

This is another large groups of bugs found in gardens, agroecosystems, woods and forests. Typical of bugs they too undergo a metamorphosis of nymphal instars and finally become winged. These are slender and mostly dull coloured. A related pyrrhocorid, is the reddish *Dysderus* bug found mainly on cotton, which has an eye catching red in its colour. In a dead specimen, one can easily distinguish it from the reduviid's or assassin bugs in that the rostrum (a long sucking tube like structure) is four segmented and not bent. Whereas in Reduviids it is three segmented. Some Coreid bugs have large legs, mainly due to expanded tibiae and tarsi. Those frequenting paddy fields and wetlands should not miss the *gundhi* bug, (*Leptocorisa*), a pest on rice. They are again lovely creatures but have an obnoxious dirty smell! (hence *gundhi*)

Leaf-footed Bug: *Acanthocoris scabrator*

25. Watch Shield Bugs

These are pentatomid bugs with a prominent and large plate like structure on the thoracic region and hence called the shield bugs. These are far smaller than the coreids. These are more rotund than slender and are brightly coloured from bluish to greenish to reddish capturing most of the hues of the colour spectrum in a rainbow! These are sap suckers and are plant feeders. They do have unpleasant odours, but quite tolerable.

Pentatomid Bug: *Andrallus spinidens*

They are harmless and can be handled. One of the commonest pentatomids is the *Nezara viridula*, which is cosmopolitan and found common in many wheat fields. Another colourful shield bug is the *Bagrada* on crucifers; which can be easily sighted in harvested cabbage or cauliflower fields; yet another is *Andrallus spinidens*.

26. Watch Tingids

Another bug commonly encountered is the Tingid or the lace-winged bug called for its lace-like pattern on their wings. These are 4-5 mm in length and both adults and nymphs are found congregating on bananas, *Lantana* etc.

Tingid Bug: *Cochlochila bullit*

27. Watch Water Bugs

One notable feature of this category of bugs is that quite a number of them are aquatic or found on the fringes of water (not marine). The first to catch anyone's attention is the giant water bug, the Belastomatid. In water, they may be mistaken for fish, and do sometimes get caught in fishermen's nets. If not handled carefully, they can inflict a sting. These are essentially buffish to brownish, about 7-10 centimetres and with a flat body profile. Watching them swim is an absorbing pastime if one is lucky enough to spot one in the edges of the water fronts. These do get attracted to light as adults, and it is then one often gets to see it. The hind legs are modified to help the bug swim like a water boat fitted with motors. A pair of apical appendages which blows air along the sides probably also help the bugs propel forwards in water. They thrive in aquatic habitats that support abundant fish spawns and tadpoles.

In one species, *Sphaerodema molestum*, the females literally molest the males! She lays her eggs on the upper side of the males which these hapless and henpecked males are forced to carry till the young ones hatch out. In India, *Lethocerus indicus* is common.

The Gerrids or water skaters are another group of bugs which can draw a lot of amount of attention. These even frequent small ponds in gardens and tanks around construction sites. They are longish and has the mid and the hind legs long (almost 2-3 times of the body length). These are perhaps the only creatures that can 'walk' on water. The wings are generally reduced as they fly less and walk (on water) more. *Gerrid spinolae* and *Halobates* sp. are quite common. They are not marine but may be found skating over calm seas, probably caused by the drifts from the backwaters.

The Notonectids or the back-swimmers are common in many small water front's like the ponds and pools. They can even gather in waters collected in drums if they have been kept too long. I have spent hours watching these boat- shaped back-swimmers moving with alacrity on the surface of water, reminiscent of swimmers who splash along belly up! While swimming on their back, their hind legs are beaten. They are equally adept in diving, most often to forage on small fish and tadpoles, using their strong rostrum (mouth parts). Generally they are about a centimeter or so in length and their "belly-up" on the water surface, easily give them away.

Another water bug, the Corixids or the water boatmen, are found submerged clinging to aquatic plants with their hind legs. An interesting feature of the bug is their ability to trap air under their wings to breathe as these lack gills like the fish. The backswimmers also trap air under their wings whenever they require to dive.

28. Watch Cicadas

Come April-May the adult Cicada bugs start their songs in Bangalore and this is the time I go out to watch them. In Bangalore, this coincides with the flowering of the Gul-mohar (*Delonix regia*), though these two have no relationship. Cicadas are found on this tree and so also on many others. These are common in many wooded tracts and forests and are more heard than seen. These are 2-3 cms in length and typically buggish. Generally the males have the sound producing organs on the abdomen, with which it gives a high pitched continuous screech. These sounds are so energy consuming that one may wonder how these might be surviving. But they do, and mate and the females lay eggs on the trees. These young nymphs which hatch do not cling to the host but drop off into the soil, where they enter into a life in the dark earth below. When the adults emerge it is promptly announced by their screeches. Their song is a little ventriloquistic, that is, not too easy to direct our attention to the direction of the sound. Moreover, with dull semi-transparent wings, they obliterate on the barks of the trees. Once sighted, the males especially are very interesting to watch. Usually they stop their music when wary of approach by anyone. If one patiently waits without movement, the Cicadas do willingly oblige with their music. It is only then one wonders how size is not limiting for high decibels. Different species of Cicadas screech different times of the year, depending on the geographical regions.

Cicadas are caught in Thailand and used as food. It is quite a delicacy. I have watched people catching them with small hand nets and transferred into jars in regions near Pattaya. There is a Megacicada in America which completes its life cycle in 13-17 years. But they do appear in predictable periods that their swarming (usually in large numbers) can be predicted. Some of the common species encountered are *Platypleura octoguttata, P. meckinoni, Pomponiya* sp. etc.

Cicada: *Platypleura octoguttata*

29. Watch Spit Bugs

Any one rambling through meadows and grasslands cannot miss blobs of "spit" on grass blades. One would easily mistake this for saliva, spat by a human. But this is the bubbly secretion of a nymph of the spit bug that lives within. Dirty your hand and examine one of these spits! You will be surprised to find a bug. The common example is the *Poophilus*. In some species like *Machaerola* the "spit" hardens into tube like structures. Usually they are dull coloured but a few bright ones also exist.

Spittle Bug: *Poophilus costalis*

These are typical sucking bugs, less than a centimeter in length, and for precision identification expert help is necesssary.

30. Watch Tree Hoppers

Tree hoppers or Membracids, are loved by ants for their sugary excretions. It is not uncommon to see several ants stroking the base of the abdomen of these insects in a quest to induce sugar flow from the anal end. This bug, a sap sucker, a little less than a centimeter has a characteristic backward bent "horn" which in fact is the backward growth of the first segment of the thorax. These are widespread in India and some of the examples found are *Oxyrachis tarandus* and *Centrotypus* sp.

Treehopper: *Centrotypus* sp.

31. Watch Hoppers, Jumping Lice and Other Plant Lice

The hoppers or jassids are the most abundant of plant sucking bugs, and occur on several economically important plants. The sizes ranges from a few millimetres to over a centimetre in some cases. The hoppers like *Cofana* are so called because of their hopping habits. They also crawl and some of them obliquely move on leaf lamina. Nymphal stages can be easily approached as they don't hop off! Many of them are dull coloured but quite some of them have greenish to yellowish hues. The hoppers are very common on mango inflorescence and an unsprayed mango tree in full bloom is the ideal place to watch hoppers like *Idioscopus, Amritodus*, etc.

Hopper: *Cofana subvirescens*

Watching Insects

On fresh mango leaves, one can see another beautiful green hopper, the *Amarasca splendens*. These have long hind legs, again adapted to hopping, but unlike grasshoppers, these are not stouter than the other pairs of legs. The hoppers belong to the family Cicadellidae. Related to these are the Fulgorids and the Delphacids which are quite common, but will need expert help to identify. A near looking group to the hoppers are the psyllids. These also have a long hind leg, adapted for jumping, hence called the 'jumping plant lice'. These are free living, but some like the *Apsylla* breed in galls in mango shoots which are common in the *terai* region of the sub-Himalayas, down to the plains of Uttar Pradesh and Bihar.

There is yet another plant lice which does not hop or jump, the aphids. These occur as winged and wingless forms, the former capable of flight and dispersal often along wind currents. These are tiny insects, may be around 2-3 mm in length, ovalish to roundish body frame with different hues of colour ranging from greenish yellow to brownish black. In most of the species, there is a characteristic pair of protrusion on the dorsal side, also called cornicles. They have the habit of living in congregation, colonizing fresh shoots, twigs and even roots. They excrete honeydew and invariably are attended by ants. While they do lay eggs, some are able to reproduce parthenogenetically. The aphids that occur on crop plants are well studied, as these also transmit plant viruses, besides sucking sap from their host-plants. Some examples are the cotton and potato aphid *Aphis gossypi*, peach aphid, *Myzus persicae*, citrus aphid, *Toxoptera* sp, etc.

Watching Insects

Potato Aphid: *Aphis gossypi*

32. Watch Whiteflies

Whitefly: *Aleurodicus dispersus*

The whiteflies are really not typical flies. Most of them are white to whitish. Some of them are even black and hence called the blackflies. These are tiny insects that belong to the family Aleurodidae. They look white because of white mealy/ powdery deposition on their wings, and are usually found in loose colonies, below leaves, or sometimes singly or in twos. The best example is the *Bemesia tabacci* which has a wide distribution across India and abroad. In Peninsular India, one cannot miss another species the spiralling whitefly, *Aleurodicus dispersus*, (which is an invasive insect into India) on account of its egg laying habits in concentric spherical circles. The nymphs which hatch remain in the spiral rings till adults emerge, which can fly. They too excrete honeydew on which sooty mould grows. The blackfly, *Aleurocanthus woglumi*, for example occurs on many fruit trees and can be seen as black, sessile nymphs of 2mm diameter, on the underside of the leaves.

Here it is pertinent to say that in most of the congregating plant sucking insects, (aphids, hoppers, mealybugs, scales and whiteflies) the excretion is sweet in the form of honey drops! These are called honeydew and are rich in carbohydrates. Ants are attracted most to those insects which colonise, as they show less movement. Some active movers like the hoppers (see under hoppers) probably distract ants and hence the latter are not seen frequenting them. Sooty mold on the honeydew is a tell-tale sign of these sucking insects as well as of scales and mealybugs which are dealt in the next chapter. Sooty mould disfigures leaf colouration to blackish and often affect photosynthesis in plants which harbour these insects.

33. Watch Scales and Mealybugs

India is famous for lac from which switches, art objects etc are made. It is of high economic value and gives sustenance to many rural folks in north India. The lac is produced by a scale insect! One characteristic of this group is its ability to adhere to one spot, singly, sparsely or in dense crowds and suck sap that moves through the host plants. Many of them are notorious pests of crops. As they hardly move, legs are atrophied. However, in mealybugs, legs can be distinguished though hardly used. Even the antennae are hardly evident. These are covered with hard waxy or powdery or mealy coating. There is another group of mealybugs, the Margarorids which have less mealy powder. So the segmented body of the insect is distinct. Some of them spend the egg stage in the soil and the later nymphal stages up the host plants.

Mealybug: *Paracoccus marginatus*

The size varies from a few millimetres to a little over a centimetre. These insects are found in every garden, woods and forests and one cannot do away with them. Some like the *Coccus viridis* which occur on citrus, coffee and sapota are greenish and get obliterated with the leaf and therefore hardly discernable. These too excrete copious honeydew attracting both ants and sooty mould. The latter is at the later stage of the insect growth and gives away the presence of the scales. However, it is difficult to eliminate them from a plant, that several of them move places with the fruit, cuttings etc.

The males are 'dipterous' in that they have only one pair of anterior wings. The hind pair is reduced considerably. In the females, wings do not develop, but they remain sessile and turn into an egg sac, releasing young immature nymphs which are called crawlers. The crawlers crawl away from the mother and adhere to new leaves/shoots or clamber to leaf edges, from where they are wind-blown to newer niches or to nowhere and die! Many crawlers adapt quickly to sap rich vegetation that these have a wide range of hosts. Once a crawler draws sap from a plant, the rostrum gets 'fixed' into the plant, drawing sap as and when needed. It is like one remaining put in a place living 24×7 with the mouth to straw in a coke bottle!

As scales and mealybugs are easily transported, people should be careful when they move plants, vegetables, fruits from one state/country to another. The best would be not to bring in plant material including seeds from other places or countries, as inadvertently or unknowingly you may introduce a mealybug or scales or any tiny insect, which may cost the country crores of rupees. One best example is the 'Papaya Mealybug' which almost destroyed the papaya industry between 2009 and 2010 when it was introduced inadvertently into our country.

The cochineal insects, famous for their dyes extracted by the cosmetic industry belongs to this group. The females of one species *Dactylopius opuntiae* are oval with reduced legs and antennae, and suck sap from cacti, and hence used as bio-control agent against weedy cacti in Australia. However the dye they give is also useful. The red scale, *Aonidiella aurantii* is common on rose and looks like red dots of pustules. Scales and mealybugs are grouped under coccids and as a group, are highly cosmopolitan. These groups also support a wide range of natural enemies at the next trophic level. However due to the high fecundity and adaptations, they remain numerically dominant.

Scales: *Diaspis echinocacti*

34. Watch Thrips

Another group of small insects mostly among flowers and foliages are the thrips. These are tiny, ranging from two to seven millimetres in length. Most of them are darkish brown to black, yellowish to orangish. These are easily seen if tapped and dislodged from flowers, tender leaves and inflorescence on to one's palm. Thrips are adapted to feeding by scrapping and sucking. Not all are plant feeders as some thrips feed on other smaller insects like aphids. One characteristic behaviour in most thrips is the upward curling of the tip of the abdomen.

Thrips: *Gynaikothrips uzeli*

The wings look 'feather-like' facilitating the interlocking of the fore and hind wings. Another trait not easily seen, but essentially typical of thrips is the asymmetry in the mouth parts. This can be seen only through a stereo binocular microscope. Thrips are not good fliers but drift through wind currents. The nymphs resemble the adults but lack the wings. A 'resting' phase before metamorphosis into adults (also called pupae) has wing pads - a fore runner to the wings in adults.

Thrips: *Frankliniella schultzei*

Some of the commonest species in India are *Scirtothrips dorsalis* and *Frankliniella schultzei* found on many vegetables like tomato, capsicum, brinjal etc. On Ficus, examination of a curled leaf would reveal the presence of *Gynaikothrips uzeli*.

35. Watch Green Lace-Wings

These are also called Chrysopids and as the name suggests are greenish. The adults are weak fliers and are about two centimeters. These generally feed on pollen. The wings are net-like and slopes side wards over the abdomen when it rests. A simple wing coupling between fore and hind wings exist in flight. The adults lay eggs on stalks, which often resemble a small bunch of erect elf-like spoons planted on leaves. In fact a female before egg laying, ejects a fluid on the leaves, which is pulled out by the abdomen to form a stalk of a centimeter long and on the tip of these, eggs are laid. The wonder here is that the thin stalk is able to support the blob of egg as the stalk stiffens. The egg stalks are anchored firmly on the leaf surface.

Green Lace-Wing: *Chrysoperla zastrowi sillemi*

On hatching, the grubs take to a carnivorous life, feeding on small sucking insects like aphids, thrips, hoppers, mealy bugs etc, Generally they are brownish to greyish with bright to black markings,. Sometimes they are mistaken for the grubs of some ladybird beetles. One can discern them by the remains of their prey which they often piggy-back on the dorsal side. *Chrysoperla* and *Mallada* are some of the common species. A near relative to the green lace-wings is the brown lace-wings which as the name suggests are brownish adults but with quite similar habits of their green cousins. However their eggs are not laid on stalks. The grubs are carnivorous on sucking insects.

36. Watch Ant-Lions

These insects are so called because of the larval forms which fondly prey on ants. The adults look delicate and resemble a dragonfly but are comparatively poor fliers. The wings are narrow and long with distinct venation. The foraging behaviour of the grubs are noteworthy, from which its name is derived. It is quite a common site to see small funnel shaped depressions of approximately three centimetre depth in less disturbed walkways and sandy patches. Beneath each funnel is a predator in waiting, with a pair of mandibles, protruding at the narrow bottom end of the funnel. Unwary ants on reaching the brink of funnel, slip into it, only to be nabbed between powerful mandibles and drained dry of its liquid contents. Other small crawling insects falling into the funnel are also eaten. The prey is usually dragged into the bottom of the funnel. Post feeding one hardly sees any debris of insect parts. Sometimes, in a square foot, more than a dozen funnels can be seen, each funnel manned by a single 'lion'. Squatting near its foraging territory is an absorbing passtime to watch these 'lions' foraging. One cannot see the ant-lion grub, but see the action of how an ant is seized, drawn further down into the funnel pit. If one is impatient to watch the hunter in action, an ant can be caught and released into the funnel. Within minutes the poor fellow is knocked off. This would remind one of a helpless deer in the jaws of a lion, which one gets to see in televisions or perhaps out in the wildlife sanctuaries.

Watching Insects

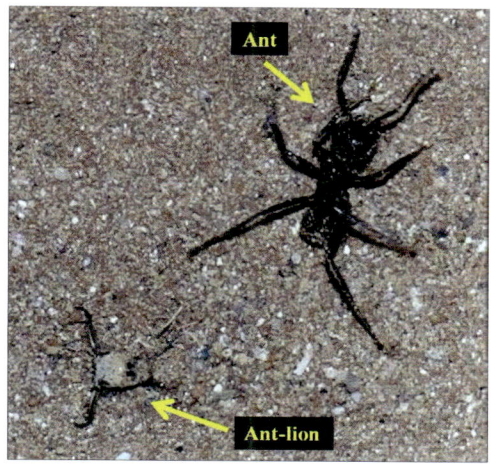

Ant-lion Preying on Ant

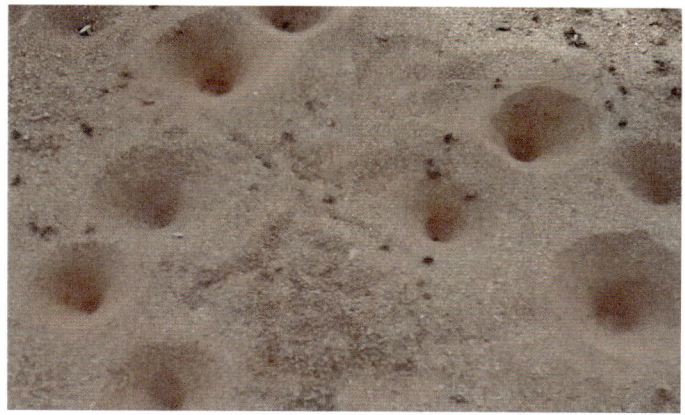

Funnel-shaped Depressions of Ant-lion

37. Watch Butterflies and Moths

The insect world is better known by butterflies than any other insects. World-over, insectaria or parks are known as butterfly parks. Butterflies are not flies nor are they buttery! The adults are seen on wings, often brightly coloured. Their counterparts are the moths, often dull coloured, but with similar characteristics. Butterflies are mostly diurnal and have a pair of antennae which end distally into a club shape. At rest, butterflies keep their wings vertical, and often vulnerable to be held by stealth catchers or even some mischief mongrels, for fun! Moths on the other hand are often nocturnal, have varied antennae (except clubbed endings) and at rest have their wings outstretched. Both these groups are called Lepidopterans, from the order Lepidoptera to which they belong. Both butterflies and moths have a totally different larval lifecycle, which pass off as caterpillars. These are often pests of several economically important crops. The caterpillars of butterflies metamorphose often into ornamented pupae called chrysalis, while the moths invariably have a dullish smooth or cocooned pupal stage. The silkworms from which the much sort after silk is made belong to Lepidoptera.

If there is a group of insects much admired even ahead of honeybees it is the butterflies. They are the winged equivalent of birds in the insect world, often rated beautiful over other insects.

Butterfly: *Pieris brassicae*

This though, is not wholly true as other insects are also beautiful. Even insect parks house butterflies largely and are called butterfly parks. Most butterflies are attracted to flowers for their nectar and some to rotten debris for fermented stuff. The nectar or fermenting juices is drawn through the proboscis which is also present in moths. Interestingly, in most butterflies the legs are good clinging agents with some even having the sense of taste on them. In other words, the foot serves as taste buds, to taste out their liquid diets. Butterflies wings bear scales and if caught while resting, these rub off on our fingers. The colour of the butterflies is due to the scales. One of the commonest butterflies is the swallowtail so called because of a tail like extension, on the hind wings, typical of swallow, a bird. The caterpillars of the swallowtails are also interesting. They are generally smooth and have a forked organ or process that protrudes and is visible when caterpillars sense alarm. These are supposed to give off repelling odours. One of the common swallowtails is the lemon butterfly. The caterpillar feeds and grows on citrus and curry leaves. The early stage of the caterpillar resembles bird dropping and gradually grows to a length of 4-5 cms. Light green and beautiful, the caterpillar slowly but surely feeds incognito on the foliage. They emerge from chrysalis as a large butterfly with a wing span of about ten centimetre. This black butterfly has pale yellow spots. In India there are more than hundred and odd swallowtails and comprise roughly ten percent of the butterfly fauna.

The largest Indian butterfly, the Southern Birdwing (*Troides minor*), with a wingspan of nearly twenty centimetres belong to this group. There are many forest dwelling species and mornings are generally the ideal time to watch them either visiting flowers or their courting or mating. The swallowtails have interesting names, probably from the British legacy and to name a few the Rose Mormon, Mime, Blue bottle, Jay, Swordtail etc. There is yet another group under them called the dragontails (*Lamproptera* spp) which have a relatively long tail and in flight resembles the dragonflies. Remember the mooncraft? These are hill swallowtails called Apollos.

Milkweeds is one group after the swallowtails that catches attention of the amateurs. The Daniads which feed on plants that on ingestion afford unpleasant taste to the butterflies or caterpillars which predators dislike. Their colours both as caterpillars and adults also catch the attention of the predators, by which they avoid the milkweeds as predators link their colour with distaste. Predators like insectivorous birds have learnt to attribute the bold colour patterns of the Daniads to distaste and hence avoid them. The Striped Tiger, (*Danaus genutia*) for example has dark black stripes on a tawny background. This is flashed around as they flit around much to the dismay of hungry birds which relish only bitterless insects!

The caterpillars of this group are no exceptions. They too have bright stripes or spots on dark or light backgrounds. Another feature is the pair of the fleshy tentacles. On pupation, usually on plants, they form a chrysalis which is supended on a thin thread. The chrysalis resembles a "heavy earring", triangularish on top and bag-like below and are usually silverish or goldenish. These are beautiful indeed to behold. The adult males usually attract the females by their attractive colours, but here the colour has a role in the anti-predation tactics. The males have a brush like organ at the top of the abdomen which releases a pheromone to attract females. The Daniads are commonly called Milkweeds (from one of their host plants), Tigers (for their tiger like striped patterns). These butterfly groups have weird names like "Tree Nymphs" and even "Crow"!

The Common Crow, certainly not the bird, is dark brownish black, wings with white spotted border and a wing expanse of ten centimetres. It is often seen in migratory congregations. In this group of Daniads, many butterflies are migratory and large swarms of butterflies moving in villages adjacent to jungles and forests and sometimes even in cities is not uncommon.

38. Watch Blues and Sulphurs

The butterflies of this group have some degree of blues on them with some whitish to grey shades. Some of the males have striking irridiscence. Like in swallowtails the hindwings have tail or long projections. Another feature here is the strong love-bond between the caterpillars and ants. The slug like caterpillar releases sugar-like fluid on which ants feed, hence the ant association. Ants however are believed to protect the caterpillar from predators, but how far this is fully true is to be explored, for most times, these caterpillars are not noticed as many of them are hidden within plants. Some bore into fruits like pomegranate and guava and come out only after pupation through an exit hole. The best example is the *Deudorix isocrates*, and another is *Chilades pandara*. The Pierrot (*Tarucus ananda*) is said to stay in the shelters built by the tree nesting ant, *Crematogaster*. Like sentries, the ants are said to guard the chrysalis, till the butterfly adults emerge. The blues are also known as Lycaenids. The Lycaenids are also known by other names like Blues, Pierrot, Grass Jewel, Pod blue, Cerulean, Oak blue, Silver Line, Yamfly, Puzzle, Cornelian, Flash, Sunbeam, Moth butterfly, Judy, etc.

The sulphurs are so called as in some species of butterflies, the wings look as though dusted with yellow sulphur powder. But they come mainly in white to yellow shades. Some of these are seen in loose migratory swarms or darting among flowers especially during bright sunshine. The Mottled emigrant (*Catopsilia pynanthey*) is

Blues: *Chilades pandava*

one of the common migrants. The caterpillars have that habit of aggregating and releasing an unpleasant odour. Fortunately, these are not easily discernable to human beings. I have seen the adults of *Pieris*, vulnerable to the bee-eaters (*Merops* sp.), an insectivorous bird.

The jezebel (*Delias eucharis*) one of the common butterflies in the Pierid group, is somewhat unique in that it has brightly coloured under wings. This catches our attention. This butterfly is said to be distasteful to its predators on account of alkaloid accumulation in its system. This group also has interesting names like Jezebel, Psych, Whites, Yellows, Pioneer, Gull, Salmon Arab, Orange Tip, Wanderer Emigrant etc. Most of these are medium sized with a wing span of 3-8 centimeters.

39. Watch Browns or Satyrids

These are one group of butterflies that eschew flowers. Quite unthinkable, as butterflies are always related to flowers. However, Satyrids prefer fallen leaf litter on wet grounds and are often found on understories of jungle floors. So, they are not high fliers and often found moving through low bushes, over the ground. At rest they camouflage with the debris of the leaves below. The adults are dark brownish to black and have noticeable eyespots which vary with species, sex, geography and season. The food unlike other butterflies is rarely nectar, but fallen fruits or logs which yield some fermented fluids. The caterpillars have two horns at the head region.

These are not sun-loving, and so like moths are active during dusk. One of the common ones is the Evening browns (*Melantis leda*) which is active at dusk and before sunrise and forages over damaged or over-ripened fruits in forests and orchards. Its caterpillar is a minor pest on rice. These have interesting names like Bushbrown, Nigger, Fourring, Fivering etc.

Paddy Butterfly: *Melantis leda*

40. Watch Brush-Footeds or Nymphalids

This is not so typical of butterflies as one would perceive a butterfly. They derive their name from the fact that the first pair of legs have tufts without claws hence brush-footed butterflies. So practically these seem four-footed at least functionally. Nevertheless, all six legs exist as they do qualify to be hexapods.

Nymphalid Butterfly: *Elymnias hypermnestra*

These four-footers are varied in colouration and sizes. The dorsal side of the wing is more beautiful when compared to the ventral. From below they look dull during flight as these are not high-flying butterflies, but while resting, their wings do catch our attention. Some of them can glide for a while without flapping their wings while in flight, hence the name Sailor is given to them. The caterpillars invariably have horns present on the head. Pupae is characteristic as they have projections. Some may have metallic colouration. Among the common Nymphalids, is the Leopard (*Phalanta phalanta*) which has a wing span of 4-5 centimeters. It is so called as the orangish brownish wings have bold dark spots and rings. They however, are also flower-visitors with an affinity to damp mud. The Pancy (*Junomia* spp) and *Euploea core* are lovely butterflies that are common even in urban areas. There are other Pancies like, the Lemon Pancy, the Chocolate Pancy- both deriving their names from our food and of course not being partial to either lemonade or chocolate! The other names in this group includes the Castor Joker, Fritillary, Painted Lady, Tortoiseshell, Eggfly, Oakfly, Map, Sailor, Sergeant, Commander etc.

Nymphalid Butterfly: *Euploea core*
(Note the short fore legs)

41. Watch Lesser Moths

Moths are an interesting group to watch. Many of the known ones are unfortunately branded as pests, hence get sprayed at with poisonous insecticides. The smallest ones are 3-4 mm in wing span and belong to family Stigmellidae. The Stigmellids are very narrow, fore and hind wings look feathery. *Nepticula* is a common moth in India. The caterpillars live mining the leaves, ie., live and feed below the epidermal layer of the leaf.

The clearwing moths are insects uncommon in the plains, but frequent hilly tracts. These are good fliers and as the name implies, wings look transparent being devoid of scales. Compared to hindwings, forewings are lengthier. The legs are hairy. The caterpillar develops inside the plant shoots into which it tunnels. Interestingly, the pupa is hooked and is capable of movement within the tunnels of the plant shoot made while it was a caterpillar. *Mellitia* is found common throughout India. Sometimes, in flight clearwings resemble bees.

There is another lesser moth, the Gelechids. These have fringed hindwings and are found in homes and storage bins. The common ones are the potato tuber moth and the grain moth.

Another lesser moth is the Gracillarids. The caterpillars are leaf miners like the Stigmellids. The resting is characteristic as the fore legs widen apart (as in stand-at-ease of the soldiers) giving the anterior part of the body a raised look. Small and beautiful, both the hind and fore wings have hairy fringes. If one has a lime plant in the garden, invariably the new leaves will be the feeding ground for the citrus Gracillarids or leaf miners. If the epidermis of a leaf is gently teased with a pin, tiny caterpillars can be seen which is *Phyllocnistes citrella*. Another leaf miner common on mango leaves and shoots is *Acrocercops* sp. There are a host of lesser moths in varied habitats and if seen and collected, expert help is required for identification.

Leaf Miner: *Acrocercops syngramma*

Leaf Miner On Mango Leaves

42. Watch Bag Worm Moths

For an eye for nature, moving bits of sticks on leaves cannot be missed. These insects construct bags with bits of small sticks (2-3 centimetres) grass bits, leaves, etc. and reside inside these bags with only the head protruding out for feeding on the leaf lamina. Sometimes, these bags are smooth and seen stuck to undersides of leaves. The bags may also be glued of small sticks/debris as in *Eumeta* spp. The bags afford excellent protection to the insect. Pupation takes place in the bags.

Bag Worm Larva In The Bag; Adult: *Eumeta variegata*

43. Watch Limacrodids

This is usually noticed when one goes into a frenzy of itching often near a tree or plant where the caterpillars are found. The caterpillars of this moth also called slug-caterpillar. These have some spiny tubercles which induce itch on touch. The adult moths are greenish or brownish with rounded wings. The caterpillar is characteristic with withdrawn legs and typical slug-movement as it feeds on leaves of trees. *Parasa lepida* is one example common in mango.

Slug Caterpillar: *Parasa lepida*

Slug Moth: *Parasa lepida*

44. Watch Wax Moths

These live in hives of honeybees, and the best known example of this family Galleridae is the wax moth, *Galleria mellonella* seen in the undomesticated bee hives.

The adult moths are sometimes called bee moths. They are distributed throughout the world. The caterpillars are buffish pink growing upto 3-4 cm feeding on the wax produced by the bees. The larvae spins silky cocoons. The life cycle is for around 7 months, from egg to senile adult. The larvae are the only ones that eat as the adults do not feed. Populations of these moths take over the honeycombs of bee colonies, usually when the bees are in a weakened state. Lesser wax moths can often be seen in bee colonies trying to lay their eggs, but in most cases the worker bees will eliminate them and keep the moths from over-running the colony. When the colony is going through a period of stress, such as after the loss of its queen bee or under starvation conditions, the moths may completely take over the honeycombs. These have their importance outside the hive also. These larvae are used extensively as live food for pets and some pet birds, mostly due to their high fat content, their ease of breeding, and their ability to survive for weeks at low temperatures. Wax worms are an ideal food for many insectivorous animals.

45. Watch Grass Moth

As the name indicates, the larvae are found in grassy meadows and also on cultivated monocots. Some of the best known *Chilo* spp and *Herpetogramma* spp are pests on sugarcane, cereals and millets. The moths are small to medium sized and rest for most part of the day. These belong to the family Crambidae.

Grass Moth Larva: *Herpetogramma stultalis*

Grass Moth Adult: *Herpetogramma stultalis*

46. Watch Flour Moth and Pod Moths

It is not uncommon in kitchens especially in villages where flour is made from pounded harvested grains on farm steads and stored in ladars. Modern packed groceries sometimes escape from their attack, but once packings are open, food products like flour, dried fruits, biscuits, broken grains, tobacco leaves, attract these moths which lay eggs into them. The larvae or caterpillar develop on these and eventually web silken cocoons as pupae. They then emerge to further colonize kitchen stores. The best examples are *Anagastra kuhniella*, *Ephestia cautella*, *E. elutella* (on tobacco leaves in stores, but not found throughout India). Though known as flour moths, many others have not made storage barns or kitchens their home. These are for example *Euzophora* sp. *Phycita* sp. and *Etiella zinckenella* found on legume pods.

Legume Pod Moth Adult: *Etiella zinckenella*

47. Watch Pyralids

These are moths that in literature get included in family Pyralidae, Pyraustidae, Crambicidae, Galleridae etc. To an amateur, these jargons are not needed as it is sufficient to know that these are interesting forms whose larvae attack several live and decaying plants.

Some of these are colourful like the *Leucinodes* which survive on brinjal fruits. Another known example is *Pempelia* sp. There are several larvae which roll leaves and feed from within (Ex: *Sylepta* sp.) The best way to detect these is to rear them till the adult emerges and get expert support.

Pyralid Moth: *Pempelia morosalis*

48. Watch Plume Moths

These are charecteristic small moths that belong to family Pterophoridae, with feathery wings and hence the name plume moths. The fore wings are so clefted that four parts are visible, and the hind wings have three parts. The fore wings are longer and the legs are slender. *Exelastis atomosa* is a common species found in India.

Plume Moth: *Exelastis atomosa*

49. Watch Silk Moths

The silk sarees we drape and displayed at window shops all come from the humble silk moth, *Bombyx mori*. This is an important industry called the Sericulture. There are several hybrids and races of the silkworm. The larvae which is whitish and smooth feeds exclusively on mulberry. The best way to understand this insect is to visit a sericulture unit or farms of sericulture.

There are other silk moths like the Saturnids, which are large with a wing span of at least 20 centimetres. One common species is the *Attacus*, which is found throughout the Orient, Africa and South America. *Antheraea paphia* is a silk moth which yields the Tassar silk. *Antheraea assamensia* yields the Muga silk in Assam. *Philosamia ricini* gives the Eri silk. Many silk moths can be sighted in forests and are very interesting to watch.

Atlas Moth: *Attacus atlas*
(The anterior end of the fore wing resembles a snake's head!)

50. Watch Tussock Moths

These are hairy moths (Family: Lymantriidae), medium sized and the caterpillars are hairy and phytophagous. The gypsy moth, the gold-tailed moth and the Nun moth belong to this group. Some of the caterpillars are beautiful like the *Euproctis* sp. The females have a tuft of hair at the tip of the abdomen, which gets detached to cover the egg mass, post egg laying. Related to these are the Tiger moths, known for their striped, coloured wing patterns on the hind wings. Moths are generally robust and most of the caterpillars are hairy.

Tussock Moth Caterpillar and Adult: *Euproctis* sp

51. Watch Noctuid Moths

Noctuid moths are known better because their caterpillars in hordes can destroy vegetations. Hence these are known as cut worms or army worms. The moths are dull brownish to blackish with spots and markings. The *Helicoverpa* and *Spodoptera* are two genera of moths which have defied agriculturists and pesticides for decades now. Another example is *Asota* sp.

Noctuid Moth: *Asota caricae*

52. Watch Hawk Moths

Hawk Moth: *Cephonodes hylas*

Hawk Moth: *Ambulyx subocellata*

This is a very interesting group of moths with a remarkably long proboscis with which it draws nectar from the tubular flowers. The antennae are generally thickened into a club and the tip is often hooked. The wings are slender and thorax and abdomen are stout. The hawk moths are good fliers but visit flowers for nectar, only in the night. The caterpillars are stoutish (resembles a small snake). *Herse convolvuli, Ambulyx subocellata* and *Acherontia styx* are common species abundant in the tropics.

Hawk Moth: *Acherontia styx*

Hawk Moth Caterpillar: *Acherontia styx*

53. Watch Flies

All flies have only one pair of wings that can be used for flying, hence the name Diptera (di=two, ptera =wings) As an insect they have two pairs but the hind pair has been reduced to a stub, hardly visible, and is called halteres. These support and lend balance during flight. All "flies" are not flies! The butterflies, dragonflies, whiteflies etc are not flies, for they have both the pairs of wings well developed. Some of them resemble a bee but can be distinguished with one pair of wings. To Diptera belongs the much disliked house fly and mosquitoes! Many flies are medically important as they spread malaria, filaria, etc. But the vast majority of dipterans are beautiful and useful as pollinators with other ecological significance.

So, flies are the most interesting of insects with a wide array of forms and colours; especially the flower visitors like the Syrphids or the hover flies. These are capable of hovering at a place in search of nectar or sites for egg-laying. These are also called the "helicopters of the insect world". Often, typical of insects, the larval forms have a different life-cycle. The maggots (as larvae of the flies are called) are predators on other insects like the hemipterans (aphids for example) with a wide array of forms and colours while many beautifully coloured adults (often with yellowish bands) are nectar feeders. Some of the maggots can be reared on rotting material. At least one of the species *Eristalis* lays eggs in aquatic surroundings.

Another familiar group is the Drosophillids. These are common in laboratories and can be bred even in homes on rotting ripe bananas. In fact, fruits in kitchens or homes attract these flies. The eyes on the sides of the heads are catchy at first sight. The fruit flies infesting guava and mangoes cannot be missed, especially the white maggots in the pulp. The maggots are vegetarians foraging on several fruits and vegetables (especially Solanum and Cucurbitae) and parts of shoots. Some also make galls on plants. They are of considerable economic importance and of quarantine significance as these move across the continents during import and export of fruits and vegetables. Hence passengers coming from other countries should avoid bringing in fresh fruits and vegetables to avoid inadvertent transfer of these insects. Once introduced, these devastate the horticultural farms. The *Ceratitis capitata*, the Mediterranean fruit fly is one such that has caused millions of dollars of losses due to high costs of its control. Fortunately, this species does not occur in India. The Indian species of common importance are the *Bactrocera* sp and *Dacus* sp of family Tephritidae. The common house fly (*Musca domestica*) also belongs to Diptera (Family: Muscidae).

Mango Fruit Fly: *Bactrocera dorsalis*

The blue bottle fly (Calliphoridae) are very common and the maggots survive on rotting materials. Calliphorids are used in forensic investigation as many of them are flesh feeders and hence useful in measuring the decay levels of rotting flesh of humans and animals. The adults like the *Chyrsomia megacephala* are weak pollinators. A peculiar fly is the Diopsid, where the sides of the head project as a pair of horns ending with the eyes.

The Hippoboscid, a flat insect is common on street dogs and cattle. These have a brownish feathery look and are about a centimetre in length.

Blue bottle Fly: *Chrysomya megacephala*

54. Watch Fleas

Fleas are not flies and they do not fly as they are wingless. Unlike the Hipppoboscids, the fleas are laterally flat. They lack a proper eye, but find their way about by walking and jumping. They are blood suckers and host on rats, squirrels, cats and dogs. When humans are attacked, they transmit plague.

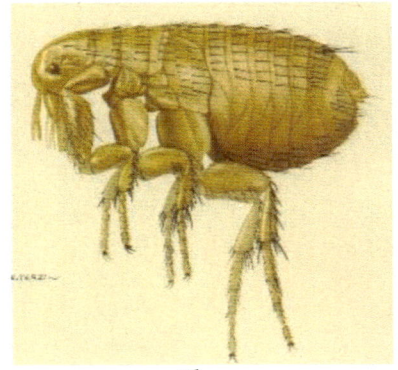

Flea

Courtesy Dr. Girish Chandra (www.iaszoology.com)

55. Watch Leaf Cutting Bees

This is quite a common bee found in gardens. The action of leaf cutting by bees is quite characteristic as it makes neat semi circular cuts on leaves. These leaf bits are carried to a nest in soil or wood. The nests are filled with these and the bees lay eggs and load it with nectar and pollen. The young grubs hatch and develop in the nest. The common genus is the *Megachile*, of the Megachilidae family. Though a bee, this is not social like the honey bee which lives in hives with a queen, workers, *et al.*

Leaf cutter Bee: *Megachile lanata*

56. Watch Wasps

Wasp: *Phimenes flavopictum*

The term wasp is sometimes loosely used for any insect that resembles a bee (Hymenopteran). The true wasps are beautiful flying insects with yellow bands or colouration on black or brownish background. The abdomen anteriorly thins into a stalk called the petiole. The most important feature is the longitudinal folding of the wings. This is characteristic of Vespidae. Like the bees, some of them have a social life with an ability to live in small colonies. Wasps have the habit of entering into homes especially on warmer days. Attempts to catch it may end up with the wasp trying to sting you. If trapped within a house, sometimes these fatigue out and drop dead. Examining a dead wasp will reveal a serration (comb-like) on the front pair of legs. These are used for cleaning the antennae. The habits are interesting to watch. They build papery, hexagonal multicelled nests (made from papery plant material) into which they place a live but anesthetised caterpillar of a butterfly or moth. The adults prey on the caterpillars and with a sting, immobilize them.

These caterpillars are then carried to the nest on which the eggs are laid. The young grubs hatch and thrive on the contents of the caterpillar. The *Vespa* sp is the one commonly called the hornet. Other species commonly found in India are the *Polistes* sp. and *Phimenes* sp.

57. Watch Beetles

The beetles form a large group in insect world. Some of the monstrous looking rhinoceros beetle or the Goliath beetle come in this group. They typically have a hard leathery fore wing (called elytra) which symmetrically cover a part of the thorax and almost the whole of the abdomen. Like hinged doors, they are held at the thorax to elegantly close and open. Below the fore wings are a pair of membranous wings folded and spread only in flight. Many beetles have well developed legs, just right for crawling. Some beetles are fond of walking that they are seen less in flight. Yet they are good fliers capable of long distance flights both vertically and horizontally. In fact many beetles can be seen atop lofty trees chewing the leaves and flowers in gardens and forests. Their sizes range from a few millimeter to several centimeters.

Beetles occupy almost all kinds of habitats. They are found on plants, on decaying matter, in soil and even in fresh waters. Like the moths, beetles have high variation in the structure of the antennae and legs. Examining an antennae with a hand lens by itself is an absorbing pastime, as mentioned in the earlier chapters.

The larval forms are called the grubs and these metamorphose into pupae which often are covered in cocoon characteristically covered by mud, debris or silk. Some pupae, however, have no covering.

58. Watch Ground Beetles

These are one of the beetles that belong to Carabidae, often encountered while crossing walkways. They are most often running. Some of them do not fly. *Anthia sexguttata*, a black beetle with yellow spots is one of the commonest species. These can be easily caught into a glass bottle for closer observations and then released. One must be prepared for the foul smelling volatiles they release.

Ground Beetle: *Anthia sexguttata*

59. Watch Water Beetles

These are long ovalish Dytacid beetles with a life adapted both to land and water. The hind legs are literally the oars of a boatman enabling the beetles to steer itself in water even in running water against a tide. Their respiratory system is well adapted to draw air from the surface of water. These have a predacious life. The eggs are laid under aquatic plants under water. The grubs on hatching are well adapted to swimming and all the three pairs of legs are just suited for that. The abdomen has a long tail propped up the water surface through which air is drawn for respiration. This is the time when it is often sighted. For pupation, these come along to the mud flats along the banks. The eggs act as paddles enabling to swim or move in water. These are shining beetles often blackish to copperish in colour. Most of them are mildly hairy with short antennae. These are also predacious. The Hydrophillids are yet another group that take to water and land. However, these are vegetarian except for the grubs. Characteristically, their antennae are short and arise from below the head. Most of them are blackish with 1-3 cm in length depending on the species.

Water Beetle: *Hydaticus vittatus*

60. Watch Rove Beetles

The rove beetles or the Staphylinids are small insects and many are less than a centimetre and easily identifiable by their mini elytra or fore wings. The hard fore wings, typical of beetles, in rove beetles, are almost cut by half. The hind wings are full but folded and tucked under the half elytra. When in need, the hind wings flip open to help the beetle take to air. The legs too are well developed, but capable of running. The beetles are ground dwelling as they feed on debris in soil and even on soil fungi and tiny insects. Most of them fit into their habitats and are sombre coloured- brownish to black, though some bright coloured ones exist. They also have a strange affinity to ants and are very interesting to watch. One should not be surprised to see the rove beetles in the company of ferocious ants and right in their nests. Some of the common rove beetles are the *Strenus, Coproporus, Paederus* and the larger *Staphylinus semiperpurus*.

Rove Beetle: *Paederus fuscipes*
Rove Beetle Preying a Bug

61. Watch Chafers

The chafers or the Scarabids are the larger beetles mostly seen attracted to light. These are a large group of beetles. They are usually a centimetre to many centimetres long and come in sombre to beautiful metallic colours. The mid-part of the thorax in some species invariably have sharp projections. They are good fliers as well as crawlers, their legs being strong. Some of the beetles have a strong projecting horn earning for themselves the name "rhinocerous beetle". The best example is *Oryctes rhinocerous* a pest on coconut and other palms. Many of these are believed to emerge after pupation in soil after the summer showers. But as a big group with varying hues and colours, they occur throughout the year in all seasons. One of the common scarabaeids is the leaf chafer, *Heterorrhina elegans.*

Leaf Chaffer: *Heterorrhina elegans*

The eggs laid in soil are often easily discernable with naked eyes. The grubs which hatch depending on the species, have long durations, lasting a couple of months. The grubs in the soil can grow up to 5-8 centimetres and are fleshy. They range from whitish to brownish colour. Often when soil is ploughed with a tractor, the grubs and for sure many soil insects are thrown up the surface, luring birds like crows and egrets that feed on them. I found that birds are attracted to the whirring sound of tractors, for the bipeds it sounds like an invite to a feast on a platter (literally). Some of the adults are also eaten up with the grubs. The white grubs as the younger forms of Scarabids are called or the root grubs are serious pests, feeding on the roots of economically important crops. The adults are also at times serious pests. Many are useful as dung feeders, for they help in detrifying huge masses of dung not only of cattles but also of forest animals.

Rhinoceros Beetle: *Oryctes rhinoceros*

62. Watch Jewel Beetles

Jewel Beetle: *Sternocera sternicornis*

These insects, as the name goes are literally gems of the insect world. The beetles are brightly and iridescently coloured. From half a centimetre to nearly four centimetres, these come in varying sizes. Most of them are greenish with a golden-orangish dash on the elytra adding on to their ornamentation. Most of the young ones of the beetles grow on trees as grubs by tunnelling into the trunk. These insects are good fliers and sometimes come to light. The species *Sternocera, Buprestis* and *Chrysabothris* are a few common ones.

> Beetles account for 25% of all known species of plants and animals. There are more kind of beetles than all plants.
> (www.si.edu)

63. Watch Fireflies

The fireflies are actually beetles. The adults emit light that is more evident in darkness. They can be better appreciated in rural and forest areas with less artificial lights. The bioluminescence are stronger in females. These, also called Lampyrids or glow-worms, have strong sexual dimorphism. The males are winged while females are wingless. The light emitted is the main intraspecific communication between sexes. Species of *Diaphanes, Luciola* and *Lamprophorus* are the common ones.

Firefly: Lampyrid

64. Watch Click Beetles

This is yet another beetle, also called Elaterids that has very acutely projecting anterior part of the thorax. The beetle has characteristically a small head which seem to be well planted into the thorax. They have an interesting click-jump sequence, which gives it its name. A lever like mechanism releases from a notch to the left of the beetle into the air. This helps the beetle to escape from predators and humans and probably the click also serves to divert attention. These beetles are also called spring beetles or snap beetles. They come in dull as well as bright colours and range from 0.5-2 centimetres in length. The grubs are slender and wire-like and hence called wireworms. The end of the abdomen ends in a hook. The common species belong to *Cardiophorus*, *Agrypnus* and *Hemiops*.

Click Beetle: *Agrypnus fuscipes*
(This is a pinned specimen; notice the pin on the right elytra or fore wing)

65. Watch Lady Bird Beetles

This is one of the lovable group of beetles. They are essentially roundish to oval with legs so concealed, that when crawling it is like a round blob, on the move or slipping past. They come in all colours and I have, after ants, spent most time watching them. From bright red to notable blue-black elytra (fore wings), they even can boast of spots and bands to differentiate among species. Polymorphs are however common. Most of them are a few millimetres to almost a centimetre.

Ladybird Beetle: *Harmonia dimidiata*

These are usually called coccinellids and are essentially carnivorous, and useful in eating up the pest insects like aphids, mealy bugs, scales etc. Both adults and grubs eat up insects. One species, *Cryptolaemus*, is widespread in India. Its grubs are capable of existing in the colonies of the mealy bugs as its most often white and resembles mealybugs (the prey). There are others like the *Harmonia, Cheilomenes, Scymnus, Coccinella* which are common.

Ladybird Beetle: *Scymnus nubilus*

The adult females lay eggs in groups. The tiny eggs are cylindrical and appear propped up. The grubs initially are gregarious but can disperse with the help of an anal foot, which it is also used as an anchor prior to pupation. Some of the adults like the *Cryptolaemus* are seen congregating on tree trunks possibly to hibernate as grubs or pupae or adults. Some species consciously avoids ants or vice versa.

Ants which tend aphids and mealy bugs for their sweet excretion, the honeydew, is said to drive away the predators. These predators also by their repulsive secretion can repel ants away. I have noticed in guava, the ant *Camponotus* and the coccinellid *Cheilomenes* excluding each other be it by whatever mechanism. This helps both the ant and the lady bird beetle to co-exist in the guava ecosystem, with aphids as a source of food for both. To this group also belong some plant-eating Coccinellids. These are *Epilachna* and *Henosepilachna*, feeding on leaves of Cucurbitceous and Solanaceous plants respectively.

Ladybird Beetle: *Cryptolaemus montrouzieri*

Hadda Beetle: *Epilachna vigintioctopunctata*

66. Watch Blister Beetles

These are fairly large and longish beetles ranging from two to three centimetres in length with striking black, red, yellow or orange patterns. Some are bluish or brownish in colour. These are fairly good fliers and dart from flower to flower, as adults are mainly flower-feeders. These, if approached or caught may excrete an oily cantharidin which can be smelly and irritative to the skin. Though the adults are phytophagus, the grubs found in the soil are carnivorous and feed on egg masses of grasshoppers and other eggs of insects found in soil. Thus they are useful though the adults are considered pests. The common species are *Mylabris* or *Zonabris*.

Blister Beetle: *Mylabris pustulata*

67. Watch Leaf Eating Beetles

The leaf eating beetles, also called Chrysomelids are widespread with maximum diversity among the beetles. They are bright metallic coloured coming in red and yellow hues. Some of them are aquatic in at least some parts of their lives. The grubs may occur in soil and feed on the roots or plant materials in soil. These beetles are beautiful and can be easily mistaken for lady bird beetles as they too have striations and spots on their elytra. However, Chrysomelids are longish and less round unlike the coccinellids. A common example is *Sagra* sp., unfortunately, many of these beautiful beetles are serious pests on crops. Some of them are slow coaches and resemble tortoise movements. So these are called tortoise beetles and an example is *Aspidiomorpha*.

Chrysomelid Beetle: *Sagra femorata*

68. Watch Longhorned or Longicorn Beetles

These are also called Cerambycids and have a characteristically long antennae which are often folded back and most of the time extends beyond the abdomen. These beetles vary in length from a centimetre to about 5-6 centimetres. Most of them are brownish with or without bright spots probably serving to deter predators. These are wood borers. Adults mostly feed on leaves or scrapes on the surface of barks or shoots. After mating, females lay eggs on cracks of barks where the grubs hatch and bore into the trees. Often wood debris or frass (woody excreta) are a tell-tale evidence of borer inside a tree.

Longhorned Beetles: *Chelidonium cinctum*

The grubs often have a protracted life cycle often extending from

4-10 months which is spent inside the tree. They also pupate within. Trees with borers often die or become weak and defoliated. One of the commonest species is *Batrocera*, found all over India. (Some of the other examples are *Chelidonium, Pseudaristobia, Aeolesthes,* etc.)

Longhorned Beetle: *Pseudaristobia octofasciculata*

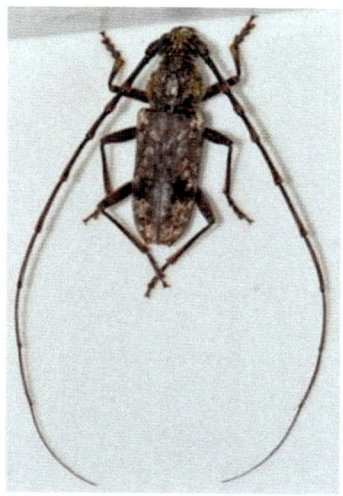

Longhorned Beetle: *Aeolesthes holosericea*

69. Watch Pulse Beetles

The pulse beetles or the Bruchids are found in dry leguminous seeds and so are serious pests on stored dals, green gram, bean seeds, etc. They are found throughout the world and billions are spent on protecting pulses from these. These are mostly brownish, about 5-6 millimetres and under a lens, show hairy covering. The eggs are laid on the pulse seeds and grubs burrow into the seed. The frass spoils the pulses. However, for an insect-watcher, an entire life cycle of a beetle can be studied in 100 gm of infested seeds of any pulse in a glass jar or a bottle whose mouth is fastened with a thin cloth. The whitish-yellow tiny eggs are sticky when laid, hence adhere to the seeds. The grubs which feed on the seed cotyledons, are curved. The adult beetles can also be found in leguminous fields mostly on the pods, as these also lay eggs in the field, which then get carried into the store houses. I used to see these often up to mid-eighties when pulses were kept open in groceries and sold by weight. These days of sachet-packed sales, the chances of Bruchid infestation have come down vastly; and we are less likely to see them on shelves and homes. Yet they are common in warehouses and seed storage bins.

Bruchid: *Callosobruchus maculatus*

70. Watch Weevils

The weevils (Curculionids) is an important group of insects that cannot escape the attention of anyone in the field. These are also called the snout- beetles. The head is extended anteriorly into a snout and characteristically the antennae is bent at right angles (or elbowed). The overall body shape is triangularly round, as the abdomen is wider than the thorax, which is wider than the head. These come in several colours- white, green, yellow, buff, brownish-black etc. The elytra has scales or mealy dustings. The legs are prominent as weevils are fond of crawling on foliages. The first part of the leg (called the femur) is swollen than the rest. The grubs of the weevils which hatch from a rather tough egg are invariably hidden in plant or soil. They are legless and soft bodied with brown head. Some can thrive in submerged aquatic plants like the water hyacinth beetles, *Neochitina*. If a clump of water hyacinth is lifted from tanks and examined grubs can be found, especially from among the drying plants.

The adult weevils with their long 'nose' are easily identifiable. Many of them, like the *Myllocerus*, feed on the margins of the leaves of plants like brinjal, mango, etc, leaving a serrated saw-like edge on the leaf. In fact, this symptom on the leaves gives the adult away. If sighted, the best is to stand and watch, for on approach, they may just disappear among the foliage. Try touching or catching one of them, they feign death and just drop down in the ground below, almost never

Weevil: *Neochetina bruchi* (with swollen femur)

Weevil: *Apoderus tranquebaricus*

to be found! Feigning death is a means of escape from predators. In tanks with the aquatic weed water hyacinth, *Neochetina* weevils can be found. Likewise *Apoderus* can be seen in mango orchards, especially in smaller trees.

71. Watch Scotylids

The bark beetles or the Scotylids are tiny weevils but do not have a long pointed snout and are tiny and are about 3-4 millimetres. As the name suggests they make tiny holes on the tree and dwell therein. Some cultivate fungi. The adults make small tunnels through the bark of trees and cultivate a fungus called the Ambrosia and feed on it. These small-holed tunnels serve as nest within which the colonies of fungus grows, on which the grubs feed.

In the seventies, bark beetles were abundant on the grape vine trunks in and around Bangalore. The tiny weevils (beetles) tunnel into the bark and push the powdered 'saw dust' backwards and hence outwards. This pasty powder sticks out through the holes and so was locally called "udubathi" meaning incense sticks as these stick-outs looked like the gathered ash at the ends of a burning incense stick!

Bark Beetle: *Hypothenemus hampei*

72. Watch Social Insects

Man is a social being and we have much to imbibe from watching social insects, especially the bees, termites and ants. I am sure for most of the times you have seen them, they are in groups both large and small and if found solitary, it is for sure that the little one is foraging or searching for food only to get back 'home' to inform of the food source.

Social insects live in groups called colonies. A bee colony, a termite colony or an ant colony can have members from a few hundreds to even thousands. It is a small kingdom for them. Needless to say, this little kingdom is ruled by the royals. In case of social insects, it's the queen who rules. Matriarchial works well in insects!

To explain how this 'ruling' happens let me introduce you to the caste system that exists in these insects. The colony is divided into three major castes. The royals, the soldiers and the workers, and as their names suggest they perform appropriate functions in the colony. The royals are the reproductives, the soldiers are at the defending and safeguarding of the colony and the workers nurse, fend and when need be, defend. Majority of the colonies also have its members looking different from the other mainly in size. The royals are the largest and winged when the mating seasons arrive. The soldiers are armoured and bigger and the workers are the smallest. On further reading it will be clear to you how this polymorphism (many-forms) varies in different insects.

73. Watch Bees

Have you ever got a compliment that you have been busy as a bee or have you ever exclaimed that you are as busy as a bee? Well, if you have, then you will know that being a bee is not easy! The life of a bee begins as an egg. The queen bee lays her first brood of eggs in a very secure and undisturbed place which she attends to as a devout mother. She is a store house of energy that she doesn't even linger out to forage. Once her first batch of daughters are active adults, all she has to do is leave the responsibility of the colony to them. for the rest of her life, she is an egg-laying machine.

Rock Bee: *Apis dorsata*

Quite obviously, a queen lives longer than the others. Why wouldn't she? She is kept at the safest and most comfort area of the hive. She is fed, cleaned, safeguarded and at times even transferred from one place to the other by the workers. All she does is lay eggs as and when she senses the need to extend her family. She does this until her capacity to lay eggs, can support to keep the colony number constant or at optimum. As the older worker or 'nurses' of the colony sense that the queen might be getting old and would not serve in laying eggs, they prepare for the 'making of a new queen'.

'What is fed to you, decides what you are going to be' in a bee colony. The usual honey and pollen makes a bee a bee. The little additionally enriched food a young larvae gets will decide its prospective future. The nurse bees give the 'destined' young ones a slight special treatment and feed with royal jelly. As the name suggests, this is only for the formation of a drone or a queen to continue the existence of a colony.

Little Bee: *Apis florea*

The eggs laid by the queen or the mother bee, are nurtured and nursed in a six sided wax cell to be what it is destined to be. It may grow up to be a drone (royal male), or might be a nurse bee or a worker/forager bee. The little egg hatches out to a small worm like larval form. It is fed with a mixture of honey and pollen called propolis. At regular intervals, it grows in size and then pupates to emerge out as a young bee.

Wood Bee: *Ceratina binghami*

Bees are also exemplary for their fidelity. They choose a host plant for collection of pollen and nectar and promptly visit and re-visit it time and again until almost all resources (pollen and nectar) have finished. By this they not only save time in searching for another plant, but also maintain uniformity and reduce contamination of pollen in their pollen stores. They pick the direction cues from the orientation of the sun and also guide the co-nest mates as to where they have to find food. Understanding their intricate dance moves earned Carl Von Frish an Austrian ethologist, the Nobel prize in the year 1973. So now you know the worth of these insects beyond monetary benefits. The bee family Apidae have several species in addition to *Apis* spp, like *Ceratina*, all of which are crucial pollinators and very much needed in our ecosystems.

Speaking of benefits, bees give us honey. Or is it apt for me to say, we take honey from the bees. The regurgitated nectar from flowers is stored as honey in the hives. With the high medicinal values it bears,

Domesticated Bee: *Apis cerana*

honey is used worldwide as a sweetener, antiseptic, antibiotic and storage medium. Also bee wax and propolis have gained importances in our daily needs. If bees were not to pollinate our crops, we may have lost at least 90% of our fruits, vegetables, oil seeds and flowers. The common large Rock bee, which builds open hives is *Apis dorsata*. Another smaller open-hived bee, common in gardens, wood lands and orchards is *Apis florea*. The domesticated honey bee, *Apis cerana* is ground hole or tree hollow nesters. This perhaps has paved way to its nesting in man-made boxed hives. Apiculture or domesticating bee is an interesting and profitable hobby. Get a training in apiculture, in a local agricultural college and start an apiary. It is rewarding these days and even *Trigona*, a stingless bee can be domesticated in a box-hive, whose honey has high medicinal value.

> Honey bees have to forage 80,000 miles to make a kilogram of honey. This may account for about 20 million trips from colony to flowers and back.

74. Watch Ants

Have you watched the movie, "A Bugs Life" ? Flik, the ant hero, Princess Atta, the grasshopper! Hope it rung a bell. Well if you have not, I suggest you do watch this animated Aesop's fable at your leisure.

The little creatures are worth watching. Sit or squat down with a cup of coffee in your garden, (if you have spared a space in your concrete jungle for those urbanised) and watch the little fellows coordinate, organise and communicate. You will be lost in a moment or two and will surely be engrossed in their busy life. What is it telling the other? Where are they all going? Why did this one get back? Oh! a trail hurdle..... you will go on. During my first ant-watch, I felt proud that I removed a twig that blocked the nest entrance of the black ant. I watched them the whole while as the ants disassembled the body parts of the hopper they had brought down home. Legs, wings, head and pieces of the body were taken in one by one in a meticulous coordination among the workers.

It is estimated that there are about 20,000 different species of ants. It is for this reason that they have been called earth's most successful species. The total biomass of all the ants on earth is astonishingly equal to the total biomass of all humans put together. Strength in numbers have gotten them this far.

Black Ant: *Camponotus compressus*

These too, like the already introduced bees, live in a colony or nest. The queen ant rules her colony and all the others you see in the nest are her daughters and sons. The ants too like the bees have a caste system, the royals, the soldiers and the workers. The queen and the winged males form the royal castes of the colony whilst the others are engaged in defence, nest building, brood nurturing or fending for food. The winged ants are the royals and the wings shed after the nuptial flight. The males or the drones have a very short life and live only to mate with the queen.

A single queen starts the colony while she lays her first clutch of eggs. Based on their habitat preference, ants make their nests on the trees, amidst leaves, in the soil, in old tree trunks, under a patch of dried leaves, under a moist rock, or even any old piece of throw away.

The queen sits and guards her eggs through the instar and pupal phases and hands over the nest activities to her first batch of daughters. The young daughters take over the task of nest expansion, foraging, nursing the young and even defence. Their co-ordination and defensiveness is well understood when you have disturbed an

Golden-abdomen Ant: *Camponotus sericeus*

ant nest. The painful sting or bites delivered will sure move you away from their nests.

Nesting Red Ant: *Oecophylla smaragdina*

Well, yes, biting and stinging are the acts of defence of ants. Sting is delivered from a needle sharp projection from the rear end of the ants body. Biting is quite literally the act of biting wherein the ant uses its razor sharp, serrated (saw like) mandible to pierce through your skin. To top with cherry on the cream, a few ants spray out formic acid along with the sting/bite and remind you of the pain for a little longer. A few such bites are quite harmless; this is only their way of asking us to keep off their territory! Why just you and me? Even huge mammals and birds are shooed away from their nesting space by these creatures.

Well developed Mandibles in the Ant, *Oecophylla smaragdina*

If you watch ants for any length of time you will see that they really do communicate with each other and very effectively too. They come in varied sizes from a few millimeters to the huger ones that can nip off your skin. Ants communicate by touching each other with their antennae. Ants also use chemicals called pheromones to leave scent trails for other ants to follow. Seeds on grass stalks, a dead insect or even a large dead animal, anything you've spilled while eating, a candy piece or the chunks of bread crumbs dropped, grains in a granary, to the decaying material on the ground can serve as food for the ants. Alarm, aggregate, defend, assemble, follow etc are all dictated in chemical language. They even transfer fluid food from one mouth and even co-ordinate and help in getting a hunt back home. To them its never the individual credit, but always the dutifulness to take the prey home and share it with the nest mates. Devout to their nests they even give up their lives fighting for its safety. Aren't you surprised already as to how the tiny little creature manages its existence? We truely have much to learn! So, watch the *Camponotus, Oecophylla, Myrmicaria, Solenopsis,* etc.

Ant-nest of *Oecophylla smaragdina*

There are as many ant species as there are birds.

75. Watch Termites

Termites are one more insect that display caste system. Termite workers and soldiers are almost always blind. Termite fathers help raise their young. You won't find deadbeat dads in the termite mound. Unlike in bee colonies and the ants, where males are short-lived and die soon after mating, the termite kings stick around. After their nuptial flight, the termite king stays with his queen, fertilizing her eggs as needed. He also shares parental duties with the queen, helping her feed their young pre-digested food. Now thats a little deviation from the usual pattern.

In almost all species, both the workers and soldiers in a given termite colony are blind. Since these industrious individuals spend their lives in the confines of the dark, damp nest, they have no need to develop functional eyes. Reproductive termites are the only termites that require eyesight, since they must fly to find mates and new nest sites.

While they can be pests to homeowners and in some crops, termites are actually beneficial insects, ecologically speaking. Termites are important decomposers. Termites break down tough plant fibres, recycling dead and decaying trees into new organic matter in soil. These hungry insects are vital to the health of our forests. As they tunnel, termites also aerate and improve the soil. It just so happens that we build our homes from termite food - wood.

Watching Insects

Its usually we who are the intruders.

Fossil studies have shown that a termite with a ruptured abdomen was encased in amber, along with the protozoans that lived in its gut. Termites digest cellulose with the help of microorganisms in their guts. Termites feed on plants directly or on fungus growing on decaying plant material. Now, that could include your home, furniture and also your stock of timber you've kept in the attic, dreaming of a warm winter. In either case, they must be able to digest tough plant fibres, or cellulose. The termite gut is loaded with microorganisms capable of breaking down cellulose. This symbiosis benefits both the termites and the microorganisms living within their insect hosts. The termites house the bacteria and protozoa, and harvest the wood. In return, the microorganisms digest the cellulose for the termites.

Termites feed on each other's faeces. Termites aren't born with all that bacteria in their gut. Before they can start the hard work of eating trees, termites must obtain a supply of microorganisms for their digestive tracts. They engage in a practice known as trophallaxis, or, in less scientific terms, they eat each other's faeces. Termites must also resupply themselves after they molt, so faeces eating is a big part of life in the termite mound. Termites are well-groomed. You wouldn't think an insect that spends its time in the dirt would be so fastidious about its grooming, but termites make an effort to stay clean. They spend a great deal of time grooming each other. Their good hygiene is important to their survival, as it keeps parasites and harmful bacteria under control within the colony.

When termite soldiers detect a threat, they send tap warning signals to the colony. To sound the alarm, soldiers bang their heads against the gallery walls to send warning vibrations throughout the colony.

Chemical cues guide most communications in the termite colony. Termites use pheromones, special chemical scents, to 'talk' to one another and control each other's behavior in a colony. Termites leave scent trails to guide other workers using special glands on their chests. Each colony produces a distinct scent, identified by a chemical on their cuticles. In some species, the queen can even control the growth and role of her young by feeding her pheromone-laden poop.

New reproductive termites kings and queens are winged, and are able to fly. These young royals, also called alates, leave their home colony and fly out in search of a mate, often in large swarms. Each royal pair of king and queen emerges from the swarm together and finds a new place to found their own colony. They break their wings off and settle down in their new home to raise their offspring.

In certain parts of rural India, during the mating season of the termites, the alates are trapped in oil plates or sweep-nets and are fried and eaten. The termite nests are broken down and the queens are hunted and also used as food. The high protein and fat content stored in the termite queen makes it mouth watering when roasted.

Watching Insects

Termites and its Nest: *Odontotermes obesus*
(Courtesy Dr. K. Srinivasa Murthy and NPIB, UAS)

The weights of all termites on earth put together is ten times more than all the weight of humans put together.

This book serves to introduce you to the insect world. It is written to suit children as well as lay public. The language is kept simple to suit school students. Though more in an Indian context, the book is relevant globally. Examples and pictures are of the Indian region, mainly from the digitals on the website of NBAIR - National Bureau of Agricultural Insect Resources, Bengaluru, India. For more on these insects visit www.nbair.res.in.

Further, surfing the net does give excellent photographs and visuals. So I advocate parallely using the internet along with this book.

Further, if insects (Entomology) have to be taught in schools, this is the ideal book.

Made in the USA
Coppell, TX
16 May 2022